## PRAISE FOR
# RISING WATER

"*Rising Waters* is solid truth in a time of denial. Journalist Tony Bartelme puts on the first-person gloves and meanders through the Lowcountry on journeys that will leave the reader sweaty, wet, mucked up in pluff mud, bug bitten and a bit befuddled. Most importantly though, is the wonder, awe, and concern for the natural world that he imparts."

**J. DREW LANHAM, PHD,**
Professor of Wildlife Ecology at Clemson University and author of
*The Home Place* and *Joy is the Justice We Give Ourselves*

"Tony Bartelme is our climate change poet, reminding us that global warming threatens not only our livelihoods but the beauty and function of nature around us. *Rising Waters* reads like the *Silent Spring* of our times."

**KERRY EMANUEL, PHD,**
Professor Emeritus of Meteorology at MIT and author of *Divine Wind*

"From a vantage point a few feet above sea level (on a good day), Tony Bartelme offers a fine view of what climate change looks like and means across the world. A valuable addition to the green bookshelf!"

**BILL MCKIBBEN,**
Author of *Here Comes the Sun* and *The End of Nature*

"With clarity of thought and passionate writing, Tony Bartelme's stories in *Rising Waters* are a call to action to protect the fragile ecosystem we depend on for so much of who we are."

**RICHARD PORCHER, PHD,**
Professor Emeritus at The Citadel and author of *Wildflowers of South Carolina*

"Tony Bartelme has an unparalleled ability to illuminate complex, global environmental problems like climate change through stories that are local, personal, fresh, and compelling."

**DANA BEACH,**
Author and founder of the South Carolina Coastal Conservation League

# RISING WATERS

# RISING WATERS

## REPORTS FROM ACROSS A RAPIDLY WARMING WORLD

"... the *Silent Spring* of our time."
**KERRY EMANUEL**
professor, author, MIT

FOUR-TIME PULITZER PRIZE FINALIST
## TONY BARTELME

Published by
**Evening Post Books**
Charleston, South Carolina
www.eveningpostbooks.com

Copyright © 2025 The Post and Courier
All rights reserved.
First edition.

Author: Tony Bartelme
Editor: Elise Lusk
Cover and layout design: Wesley Strickland

No part of this book may be reproduced or transmitted in any form or by any means, electronic or mechanical, including photocopying, recording or by information storage and retrieval system – except by a reviewer who may quote brief passages in a review to be printed in a magazine, newspaper, or on the web – without permission in writing from the publisher. For information, please contact the publisher.

First printing 2025.
Printed in the United States of America.

A CIP catalog record for this book has been applied for from the Library of Congress.

ISBN: 979-8-9909493-7-9 (Paperback)

Cover photography: Jared Bramblett

*For Margaret Bartelme,*

*Mom, teacher.*

**The Post and Courier**

# RISING WATERS
―――――― CLIMATE STORIES OF THE SOUTH ――――――

A portion of the proceeds from the sale of
*Rising Waters: Reports from Across a Rapidly Warming World*
supports the Rising Waters Lab.

The Rising Waters Lab is a community-funded project at *The Post and Courier* dedicated to sharing stories about how climate change and flooding impact South Carolina's diverse communities. You, your family, neighbors, and friends are the driving force behind this project.

# STORIES

---

| | |
|---|---|
| SNAKES | 1 |
| CHASING CARBON | 5 |
| EVERY OTHER BREATH | 15 |
| FADE TO WHITE | 33 |
| SCUM | 45 |
| LOWCOUNTRY ON THE EDGE | 57 |
| INTO THE GULF STREAM | 71 |
| THE GREENLAND CONNECTION | 89 |
| GHOST BIRD | 107 |
| OUR SECRET DELTA | 123 |
| THE SAHARA CONNECTION | 141 |
| RISING WATERS | 155 |
| WHY THIS BOOK? AND THANK-YOUS | 165 |
| ABOUT THE AUTHOR | 169 |
| ABOUT THE PHOTOGRAPHERS | 171 |
| NOTES ON SOURCES | 173 |

# SNAKES

IMAGINE YOU'RE WALKING DOWN A PATH. IT'S A BEAUTIFUL DAY. Blue skies. Light is streaming through the trees. Life is good! Until something catches your eye. Down by your feet – a poisonous snake. Coiled and ready to strike. Look at its fangs.

In that moment, what are you *not* going to do?

Deny its existence?

Check your Instagram?

No, I guarantee you that your frontal cortex will light up. Your brain's official freak out center, the amygdalae, will go berserk. And in a flash your brain will tell you to get the bleep out of the way of that snake.

Our brains are wired to respond to visible and immediate threats. And that's a problem when it comes to global warming.

Carbon dioxide, a major greenhouse gas, has no color, no odor, no threatening hiss.

It's an invisible snake.

Melting ice in Greenland? It's so far away, we don't see or hear the hiss of its cracking ice.

As a result, our brains process these and other hidden climate threats with much less neural drama. Those amygdalae remain on standby. Networks that flush adrenaline into your bloodstream stay dark. This creates an opening for other thoughts. Should I buy a new couch? The thing the politician just did – uncool! Distractions pile up in our minds like unread emails.

So, no wonder we're not moving out of the way of a rapidly warming planet. Our neural wiring is biased in favor of what we see in front of our faces. We're only human.

But what if we could see these invisible snakes? What if we could see carbon dioxide? See those changes in ocean currents? See those microscopic oxygen producers hiding in the waves? What if we wake up those amygdalae in our noggins?

Hold onto those thoughts. We're going to do a little rewiring in this book – make the invisible visible.

But first, let's talk about waders.

---

I live in an old neighborhood in Charleston, South Carolina, on a street that often floods. When the moon is closer to Earth, and the sun's position is just right, the extra gravitational tug makes our tides a foot or two higher. We call them "king tides" or "sunny day tides." When they arrive, seawater oozes from the ground like a stepped-on sponge. The Atlantic rises through storm sewers and pools by the curbs. We've always had king tides, but they're different today. Four decades ago, low-lying areas of downtown Charleston flooded once or twice a year. Now, it happens seventy times or more.

Part of me loves these tidal inundations. They're reminders of the sea's power, lessons on the inescapability of change. But rain tests this love. Add a downpour to a high tide, and my street turns into a knee-deep river. During one flood, a construction crew's porta-potty washed away. I've seen car bumpers float by. I've kayaked around the block without touching the ground. I watched someone try to windsurf in a park a few blocks away. I've found fish swimming at the foot of my front steps. During downpours, my neighbors scramble like a volunteer fire brigade to move their cars to higher ground. One night I slept through a storm, and the next morning the water was up to the top of my car's stick shift. That car forever smelled like a Lowcountry marsh.

Waders help – they give you freedom to walk through this murky mess. Mine are ugly brown things made of rubber that reach up to my crotch. In some floods, they weren't high enough and fetid water poured in and pooled around my toes. But most of the time, they work just fine. In fact, I often feel invincible in them. Protected, I can interview motorists who just abandoned their vehicles. I can trudge through the water to the drug store as if nothing happened. I once carried a date on my back to a Christmas party. We've been together since, and she has her own waders now.

Waders aside, Charleston is the perfect place to investigate climate change issues. Much of the city is at or just above sea level, so every inch of sea rise remakes the area's edges between land and water. Edges are where the action is, whether it's a front line in a war or a beachfront home suddenly surrounded by water. New edges create tension, conflict. And conflict is at the heart of any good news story.

And stories are important. Words are basically triggers for neural flashes in our brains. When you herd these neural cats into stories, you're working brain to brain, writer to reader. And stories are how human beings have always learned lessons. They help us get out of the way of snakes, visible and hidden.

The stories that follow grew from work I did for *The Post and Courier*, a locally owned newspaper that has a national reputation for doing deep dives. Going beyond the obvious – that's always been my goal. Look below the surface to problems that have been cloaked by complexity and misinformation; identify connections that major media outlets missed; search for moments when scientists had aha moments. This usually leads you to deeper themes, ones about loss, redemption and beauty.

Experienced guides are important in journeys into the unknown. And I've met fascinating ones doing this work: a NASA scientist who moonlights as an Elvis impersonator; a botanist nicknamed "Hoot Owl"; an Inuit shaman; and an argonaut who rode on a real Yellow Submarine.

And those invisible snakes?

Let's look at some of them now.

# CHASING CARBON

IT'S JULY IN DOWNTOWN CHARLESTON, AND THE NOON SUN feels like a Bunsen burner. I'm in a parking lot near a busy intersection off King Street and holding a rare camera. I'm a little nervous. The camera belongs to a company called FLIR. "Our engineers built it as a technical challenge, and it's the only one in the United States," the FLIR executive had told me. "It cost something like ninety thousand dollars, so don't break it." And then he'd FedExed it to the newsroom.

FedExed a ninety-thousand-dollar camera!

At first glance, the camera looks like one of those bulky videocassette recorders back in the 1990s, something you might find in a drawer with your Sony Walkman. But hidden inside this camera is a lens made with a rare metal called germanium and a tiny freezer that cools an infrared sensor. Once cooled, the sensor captures wavelengths absorbed and emitted by carbon dioxide. Using this data, the camera generates images of carbon dioxide plumes. *Voila*, the snakes.

The first target: my car's tailpipe. I set up a tripod and put the camera on it. It's windy and the camera wobbles. I point it toward the car's rear bumper. People walk by with quizzical looks. I rush to the driver's seat, start the engine, half-expecting to hear the camera behind me fall and burst into bits of germanium and ice. I rush back. It's still standing, phew. I press the camera's buttons, not entirely sure what all of them do. The camera groans. My mind conjures an image of a tiny freezer blowing up.

It takes a few minutes for the camera to warm up, or rather, cool down. While waiting, I remember something from my childhood in the early 1970s. We lived in the Los Angeles area, and on some days, the smog was so bad that we couldn't see across the street. You could feel your lungs burn after playing outside for a few minutes. But tough pollution regulations forced car makers to innovate and cut soot and other visible pollutants. In the 1970s, Los Angeles had unhealthy air five days a week on average. Now it's about two days a week. Progress.

But none of these regulations targeted carbon dioxide emissions.

As I stare at the car's tailpipe, I see no visible smoke. Nothing. Then the camera stops groaning. Ready.

I look at the viewfinder and see a different world.

Grainy images show bright plumes of carbon dioxide shooting from the tailpipe. I play with the camera settings, changing these images from black and white to color. I like the ones in color the most. The carbon dioxide billows out in reds and oranges instead of smoky grays. I switch the settings so these clouds of carbon dioxide look beautiful. It's hypnotic. Like watching a campfire.

Next, I point the camera toward cars and trucks as they roll down King Street. I notice their plumes have different shapes and sizes. Plumes from newer and smaller cars are relatively small, but the SUVs and trucks are much larger. Some look like flame-throwers, which makes sense. The average passenger car produces about ten thousand pounds of carbon dioxide a year. But an SUV or pickup generates 13,600 pounds—about 40 percent more than cars.

What about larger vehicles? A busy bus stop is four blocks away, so I set up there. I watch buses come and go as the camera reboots. Some of the buses are packed, but most are empty. Empty or full, the buses spew much more carbon dioxide than the SUVs and cars. As I fiddle with the camera settings, a bicycle rider passes an idling bus. No visible carbon dioxide from the rider. Tons of it spewing from that stationary bus.

A special camera shows an idling bus pumping out CO2.
Photo by Tony Bartelme of *The Post and Courier*.

I think about mass transit for a moment. It's supposed to be a greener form of travel. But what if those buses are empty? Could a lightly used mass transit system be worse for the environment?

While mulling this, a tractor-trailer rolls by with a carbon dioxide column rising behind the cab. Each day on average, the United States truck fleet generates 2.2 billion pounds of carbon dioxide. All told, internal combustion engines generate a third of the world's greenhouse gases.

Because it's so hot, I head back to the newsroom and camp out for few minutes, thankful for the air conditioning. A few cubicles away, I hear familiar voices. Brian Hicks, a columnist, is talking to another colleague. I grab the camera and begin rolling. Brian can spin a yarn, so I figure that he's also a significant source of carbon dioxide. But the viewfinder shows just a few wisps from his mouth. Not as much hot air as I thought. Still, wouldn't eight billion Brians add up? After all, we breathe in oxygen, and we burn fats, proteins, and carbs instead of gasoline. And then we exhale carbon dioxide. We're all engines.

The good news is that most of our food comes from plants, which take in carbon dioxide and sequester that carbon in pistachios, tomatoes, and other delicious plants. We eat animals, of course, but move down the food chain, and you eventually find ones that eat plants. So, whether we're eating spinach or turkey bacon, our food ultimately is rooted in photosynthesis. Plants that take in carbon dioxide essentially cancel all that carbon dioxide humans exhale. It's called a closed-loop cycle. Bottom line when it comes to carbon dioxide, eight billion people breathing out $CO_2$ isn't a problem.

The bad news is that eight billion people also use a lot of internal combustion engines and electricity, and much of that energy comes from coal and natural gas. That air conditioning in the newsroom? Burning coal makes those compressors and fans hum. We burn to cool.

Does all this burning really turn our atmosphere into a greenhouse?

Here's it's helpful to step back in time.

---

In 1896, a Swedish scientist named Svante Arrhenius published a paper with a bold hypothesis: burning coal releases carbon dioxide, which creates an atmosphere like air inside the "glass of a hothouse." Perhaps because he was from Sweden, Arrhenius said that a warmer planet might be a good thing. He speculated that an increase in carbon dioxide and a balmier climate would stimulate plant growth and provide more food to a growing population.

Arrhenius was the first to propose that human beings could affect the climate's temperature by burning coal. And for the next eighty years, climate scientists didn't worry too much about the greenhouse effect. We measure carbon dioxide in parts per million. Over most of the past eight hundred thousand years, the atmosphere's concentration typically hovered around two hundred parts per million. Occasionally, it exceeded three hundred parts per million. In the late 1950s, scientists began regular monitoring on top of a volcano in Hawaii and pegged levels at 315 parts per million.

Scientists began ringing the alarm bell a little louder in the 1970s. In 1981, a team led by NASA scientist James Hansen found that global temperature had already risen about three-quarters of a degree that century, with half of that increase happening in the previous twenty years. He and his colleagues pointed at the rapidly rising levels of carbon dioxide from coal and petroleum products. That year, carbon dioxide levels on that volcano in Hawaii hit 340 parts per million. At 350 parts per million, the climate heats up so much that the polar ice caps melt.

In the newsroom, I ask Google to find the latest carbon dioxide measurements from that station on the Hawaii volcano. The answer arrives with a jolt – more than 420 parts per million. The atmosphere hasn't held this much carbon dioxide for four million years.

---

Blinders off, I hunt for more sources of carbon dioxide: a train at the South Carolina Ports Authority shipping terminal; a ship underneath a bridge; planes at the airport; and perhaps the best and easiest thing humanity can get rid of—gas-powered leaf blowers. Aside from their ear-splitting noise, gas-powered leaf blowers are among our worst polluters. Gas-powered leaf blowers and lawn equipment often use two-stroke engines, which mix fuel and air with lubricants and generate a soup of toxic chemicals and carbon dioxide. By one estimate, using a gas-powered leaf blower for an hour generates the same amount of pollution as driving a Toyota Camry from Charleston to Chicago. These small sources make large impacts when you add them up, but what about larger sources?

I jump into my gas-burning car and unlock carbon dioxide for forty-five minutes on my drive to the Cross Station, the largest coal-fired power plant in South Carolina.

The Cross Station sits on the edge of Lake Moultrie. I stop on a bridge over the railroad tracks and set up the tripod. Below, a coal train stretches from the plant toward the horizon, its cars full of black chunks glinting in the sun. Cross Station's stacks rise higher than a forty-story building, casting shadows on hills of ink-black coal fed by those rail cars. Just before I turn on the camera, a man on a Harley passes by, makes a U-turn and roars back toward me. He stops a few feet away and in a curt voice asks me what I'm doing. I tell him about the camera. He says he works at the plant. I try to make small talk, but he scowls, turns his bike around, and roars off in a cloud of smoke and carbon dioxide.

I turn back toward the plant. Every day, conveyors and cranes feed this coal into crushers and send the black powder to furnaces that burn at 2,500 degrees Fahrenheit. Heat from the furnaces turns water into steam that spin four giant turbines. Those turbines generate 2,390 megawatts of electricity, enough to power more than 1.1 million homes.

Big white clouds of steam rise from its stacks. They're pretty, like regular clouds. But when I look at the camera's screen, I see the carbon dioxide. Big masses of gas, rolling into the sky.

In 2007, I was in the office of Lonnie Carter, chief executive officer of Santee Cooper, the utility that operates Cross Station. I was working on a story about the utility's push to build another billion-dollar coal plant near a remote swamp in South Carolina's Pee Dee region. I asked Carter about climate change. He'd told me that he thought it was "a little bit gaudy to think that we can actually affect the climate." Besides, the United States was the "Saudi Arabia of coal," he'd said. "We can't abandon coal."

But after our conversation, conservation groups challenged the utility's argument that it needed a new coal plant. Santee Cooper eventually backed down, but not because it agreed with the environmentalists. Instead, electric utilities like Santee Cooper found something cheaper to burn. The fracking industry had become so efficient that the United States had become the "Saudi Arabia of methane." Natural gas prices had fallen. Coal plants had become more expensive to operate, and utilities had begun to mothball their older coal burners. All this came with what seemed like a climate bonus: natural gas plants release less carbon dioxide. The industry's marketing machine began pitching them as "clean burning" plants that were "bridges to a greener tomorrow."

I wondered what a clean-burning natural gas plant might look like with the FLIR camera, so I headed back to Charleston. The oldest part of the city is on a peninsula with a "neck" that connects it to North Charleston. On this neck, a utility built a "peaker," a power generator it can crank up during periods of peak demand, such as hot days like this. And it runs on natural gas.

When I arrive, I see nothing from its stack. Surprising, given that the heat index is nearing one hundred. Then I look through the viewfinder.

I see a blast of carbon dioxide coming from its stack—a giant carbon dioxide-spewing torch.

Later, I look up the plant's emission data. Largely unseen, this natural gas plant pumped out about 48.3 million pounds of carbon dioxide a year.

By the end of the day, I'm soaked in sweat. This month will turn out to be the hottest July in Charleston's recorded history. But I'm oddly energized by this new way of seeing the world. All around me are invisible fires. Knowing this is helpful. If we want to reduce global warming, we need to become better firefighters. Simple as that.

My time with the camera is up, its images branded in my brain's neural wiring. I drop off the camera at the FedEx distribution center. The camera's journey back to a FLIR office in New Hampshire will generate its own carbon footprint, its own fire. One that I won't see. So maybe it's not as simple as that.

At least I'm thinking about heat in a new way, I tell myself. And my mind drifts toward another question. Where does all this trapped heat go?

I come across a calculation from two scientists: every second, human beings create greenhouse gases that trap heat in the atmosphere equivalent to four Hiroshima-sized atomic bombs.

Four nukes a second.

And 90 percent of that heat goes into the ocean.

So that's where I'd look next—climate stories hidden in the waves.

# EVERY OTHER BREATH

IT'S EARLY IN WORLD WAR II, AND AMERICA'S FATE IS TIED TO the sea. In the Atlantic, Nazi U-boats pick off convoy ships, threatening the lifeline to Europe. In the Pacific, Japanese subs prowl the depths, testing a fleet battered at Pearl Harbor. Meanwhile, American war planners know little about this new underwater battlefield. Mastering its terrain could prove decisive. In a secret report, the leader of a hastily assembled group of scientists called Division 6 writes about an urgent need to understand the ocean "for service in a national emergency."

New sonar instruments would be the tool. You could ping the ocean bed, then use the bounce to calculate the seafloor's depth and shape. But sonar operators soon pick up something odd: the seabed seems to move up and down. More pings, and Navy technicians uncover a pattern: this "false bottom" rises at night and descends at dawn.

What is it?

Must be alive, the Division 6 scientists think. Squids? Schools of fish? Must be something pervasive; the sonar picks up false bottoms in all of the world's oceans. Whatever it is, could American submarines hide under it? Could enemy subs?

Division 6 doesn't answer these questions before the war's end, which only stokes more curiosity. Scientists begin calling the false bottom the "deep scattering layer." They toss nets into the layer; they haul up a few squids and fish but not enough to explain those scattering pings. Then, with fine mesh nets and deep-sea diving gear, scientists in the 1970s finally solve the mystery: the false bottom is a massive daily migration of plankton.

This symphony of tiny and beautiful creatures begins at night when they rise to feed on even smaller surface plankton. Countless fish join this movement—so many that the ocean hums. Then it ends at sunrise as they plunge to escape predators. Though unseen, this daily cycle is the grandest migration on Earth. From an ecological standpoint, it's exponentially more significant than the Serengeti's thundering wildebeest or the winged journeys of the world's birds. And it's just a small part of the plankton story. Startling new discoveries about plankton could prove decisive in an emergency that's as urgent as any war: a rapidly changing climate.

The question remains: will we learn enough in time?

---

Plankton was the most important stuff I'd never heard of—more important in some ways than the rainforests, at least when it comes to what we breathe. But I wasn't thinking about climate change when I stumbled into this hidden world. I was just trying to regain my balance.

In 2015, Charleston was the scene of two horrifying national stories: a white police officer pumped eight bullets into the back of a Black man named Walter Scott after a traffic stop and chase.

A few months later, a white supremacist killed nine Black people after a Bible study at Mother Emmanuel AME church. Journalists from around the world converged on the city for both stories. Our newsroom worked at full throttle week after week. After the satellite trucks were gone and the funerals were over, we were all out of breath. Then I came across a book that took my mind off the collective grief of the city.

Book publishers send newspapers stacks of books every week, hoping for reviews. In our newsroom, most ended up on a counter in the corner of our library. Passing through the library one day, I spotted a coffee-table book with an eye-popping cover. From a distance, it looked like a photo from the Hubble telescope, a kaleidoscope of colors set against a black background. Its title was *Plankton: Wonders of the Drifting World* by the French biologist Christian Sardet. I opened it and was transfixed. It had bacteria that looked like microscopic planets. The algae looked like jewels. Sardet had photographed many of these drifters with black backgrounds, bringing out their shapes and colors. The beauty of these images was soothing amid the wreckage of what I'd been working on. I wanted to learn more.

The term plankton is a catchall of sorts for living things that are at the mercy of currents. That broad definition includes jellyfish, krill, marine bacteria, viruses, algae and fish larvae. With no roots to the seabed, planktonic creatures mostly drift, though as those World War II pings revealed, some are masters at moving up and down. Many forms of plankton are so tiny you need a microscope to see them, but they are the unsung heroes of the planet's air.

You can thank species of sun-loving plankton for the breath you just took. Until about two billion years ago, the planet's atmosphere was breathlessly devoid of oxygen. But then a distant cousin of today's blue-green algae began using the sun's rays to split water into hydrogen and oxygen. Earth hasn't been the same since.

Today, half of the atmosphere's oxygen comes from ocean plankton—every other breath. Plankton comes in all shapes and sizes, but scientists divide them into two categories.

Phytoplankton are the microscopic algae and other cells that drift in the sun-infused upper layer of the ocean. Think of them as the plants of the sea, the oxygen producers. Zooplankton are the larger animals that typically feed on the phytoplankton. Think of them as the sea's insects, snails, and worms.

Small and large, plant and animal, they do amazing things.

One zooplankton called phronima chomps on smaller plankton and uses their body parts to make protective cellulose barrels. The mother lives in the barrel, zealously guarding her young, while males bolt at the slightest danger. Phronima's fierce appearance was the inspiration for the creature in the 1979 movie *Alien*.

Some algae and bacteria have red pigments and grow so dense they color vast areas of the ocean; floating pink blankets of cyanobacteria gave the Red Sea its name. One zooplankton species has little blue sails; washing up on the beach, their colonies look like a regatta of tiny blue boats. Jellyfish also are in the plankton family, despite the "fish" in their name. They're actually invertebrates that drift with other plankton. One jellyfish—the Turritopsis dohrnii—can literally reverse its aging process, moving from its adult stage back to a polyp, then maturing again. In the absence of predators, it could hypothetically repeat this cycle forever. Its nickname is "the immortal jellyfish."

When phytoplankton die, they emit cloud-forming chemicals that give beaches their intoxicating and briny smell. Some phytoplankton also are killers: shaped like glistening needles, they secrete neurotoxins that find their way into shellfish and animals that feed on them. After one toxic algal bloom in 1961 in California's Monterey Bay, thousands of poisoned seabirds dive-bombed houses and cars and piled up dead on streets. The phenomenon inspired Alfred Hitchcock's 1963 movie *The Birds*.

Counting plankton is like counting stars, though instead of stars, some species look like tiny Christmas ornaments and the Leaning Tower of Pisa. A single teaspoon of seawater might contain one million phytoplankton bacteria and one hundred million planktonic viruses that feed on them. Viruses alone have an overall biomass in the oceans of seventy-five million blue whales. Mostly unseen, these hordes of planktonic algae and viruses are in a constant state of biochemical warfare: the viruses attack the algae while the algae develop special plates of armor.

Much of what we know about plankton has been discovered since World War II, thanks in part to those curious Division 6 scientists, but also because of people like Dennis Allen, who I met one morning on a boat in a South Carolina marsh, as he wondered why the water looked so weird.

---

Dennis Allen was resident director of the University of South Carolina's Belle W. Baruch Marine Field Laboratory, a science center deep in an area of protected pinelands and marsh north of the small city of Georgetown. He has a boat captain's beard, full and white, and his brown eyes hold a look of delight. He has long been fascinated by the hidden underwater world, and it's easy to imagine that delighted look sixty years before when he watched his Amazing Live Sea Monkeys hatch. He'd gotten the kit, which advertised how you could: "Own a bowl full of happiness! Because they are so full of tricks, you'll never tire of watching them."

Those brown bits didn't really look like monkeys, though. And Dennis would learn much later that they were really brine or fairy shrimp. But they darted and moved about nevertheless, a small world of their own that hinted at a much larger one if you just looked closer.

Look deeply at things—that's what Pop-Pop, his grandfather, also taught him. Pop-Pop was an Italian immigrant and successful auto dealer. He and Pop-Pop often went fishing off the New Jersey coast. His grandfather seemed to be curious about everything in the water, which was infectious to an already curious child.

"Pop-Pop, why do we catch flounder here but not over there?" he asked his grandfather.

"That's a good question!"

"Pop-Pop, what do the fish eat?"

"Let's see."

Gutting the fish, they found dead shrimplike animals in the stomachs—not so different than those Amazing Live Sea Monkeys.

Fast-forward to the mid-1970s. Dennis was a biologist by then, working on his dissertation, trawling with a fine mesh net off New Jersey, funneling zooplankton into a jar at the end of the net. He pulled up the net and looked at the jar. It was full of mysids, those shrimp-like animals he saw when gutting fish with his grandfather. From a distance, the jar looked as if packed with wild rice. But looking closer, you could see it teemed with life. New questions formed in his mind as he watched the mysids flit back and forth: what is their response to light? How do they move with the currents? Years later, he still has the jar on a shelf at the Baruch

Institute. He smiles. "It's a reminder of my roots"—as well as the discovery that fueled his most important work.

In January 1981, not long after he landed a job at Baruch, he collected samples in the salt marsh creeks of North Inlet. He recorded the water's temperature, salinity, and other chemical characteristics. He collected the zooplankton with a fine mesh net.

Then he kept at it, sampling every two weeks, using the same nets and protocols.

And he kept going, even when the funding was barely there.

As the decades passed, he and his colleagues filled shelves with glass jars of the samples, preserved in a pink fluid; data filled columns and charts. Like a savings account, it slowly grew in value. Together, the data makes up what scientists call a "time series," and Baruch's is the longest continuous time series in an estuary in North and South America, and perhaps the world.

The rarity and value of a good time series can't be overstated. With data collected consistently over a long period, scientists begin to get a sense of long-term trends. In this case, Dennis had learned how estuaries and salt marshes change over time.

And on the boat in January, he told me he was shocked by what he saw.

---

It's the 866th collection in the time series, and the first of 2016. The marsh is a winter palate of pale yellows and greens. A bald eagle watches from the top branch of a dead tree. The boat moves into the current. The water is brown, the color of iced tea, and this is news.

"This time of year, it's supposed to be grayish-green and really clear," Dennis says, as Paul Kenny steers the boat. Kenny, a research specialist, has been collecting samples with Dennis for thirty-three years.

The source of the unusual color is last fall's torrential rains. A record two feet fell in one weekend. It's the kind of rain bomb scientists think we'll see more of as the planet warms and the air holds more moisture. More rain fell in November, fueled by one of the largest El Niños climate patterns in recent history. This morning, salt marshes have the salinity of freshwater ponds.

Strange weather, but Allen's time series has revealed other surprises that go far beyond any blips in the jet streams. He grabs a net that looks like a windsock and tosses it overboard. The net funnels zooplankton into a jar. He and Paul pull in the trawl, and Dennis lifts the jar to the sky. A translucent eel larva flickers amid the debris. It probably migrated from the deep ocean, perhaps from the Sargasso Sea.

"The big story is that there are a lot fewer zooplankton in North Inlet than there used to be," Dennis says.

When he began in 1981, a trash-can-sized amount of water held ten thousand to twelve thousand zooplankton specimens.

The numbers declined over time. In recent years, they've been catching about six thousand to eight thousand.

"That's a 40 percent reduction," he says. "That's huge, and it's remarkable because it happened in just the course of thirty years."

Rising ocean temperatures might be responsible, he says.

And all these changes affect the mostly hidden world of plankton, the foundation of the food web.

Dennis says they once caught hundreds of anchovies in trawls like today's. But in recent years, they've caught one or two per tow, or none at all. Researchers in the Chesapeake Bay, the West Coast,

New England, and Europe are discovering similar reductions in zooplankton and fish species, he adds.

Will this affect the climate?

And the air we breathe?

After my visit with Dennis and Paul at Baruch, I set off for Bermuda in search of more clues.

---

Across the Sargasso Sea, three young scientists on the boat *Rumline* prepare a gray canister that looks like a small torpedo. In the distance, I spot the pastel buildings along Bermuda's rocky coast. The boat slows. White foam breaks over a nearby coral reef. A Brazilian technician lowers the canister into gentle waves of turquoise. He hoists the canister back on board, and the three scientists take turns piping the contents into beakers. They repeat this four times at other spots off the island, then give a thumbs up to the captain, who turns the boat, painting a white arc of bubbles in the blue as it heads back to the dock at the Bermuda Institute of Ocean Sciences. In a lab there, they'll take those beakers of seawater and measure salinity, pH and other aspects of the water's chemistry. It's work like Dennis Allen's time series in South Carolina. But instead of estuaries, the Bermuda scientists focus on changes in the open ocean. As far as time series go, Bermuda's is among the longest-running ones in an ocean setting, and, therefore, among the most important.

I met Amy Maas, a scientist at the institute, who explained why.

"People always ask, 'Why should I care about what's going on in the open ocean?'" Her eyes open wide, and her hands dart like fish. "And I say, 'It's all connected! Our water! Our air! And this place is one of the key places in the world where we've learned about the importance of marine bacteria, how migrating plankton change things, how the carbon cycle works."

Growing up in Ohio, Amy thought she would be a librarian. But the more she learned about marine biology in college, the more she found herself captivated by its complexity and beauty. Drifting like plankton from course to course, and then in expeditions to Antarctica and the North Pacific, she was pulled toward studies about climate change and a zooplankton group called sea butterflies, also known as pteropods. Now, she works in Bermuda with her husband, Leo, a Spanish zooplankton researcher, often bringing their new baby, Bastian, who sleeps in a crib next to her desk.

As we talk, a storm lashes the island, turning the cobalt waters below her office window into froth. Winds whip around the building, generating a sound like Tarzan's jungle yell. Amy gains steam as she explains her work, sentences tumbling out in excitable bursts. Her voice rises and falls as she recalls scuba diving with creatures from the deep scattering layer, the humming migration of plankton and fish that baffled the World War II sonar operators.

"During the day, you might see a fish here and there. But at night, when everything rises to the surface, boom, everything is there! The water is packed full of stuff. Wow. It's extremely dark, and you can't see anything beyond your big light. It looks like confetti, but it's all moving around. It's vibrant. You have these flashes of color. Everyone's trying to eat and mate and do everything that needs to happen. It feels as intense as a coral reef or a rainforest. But everything is tiny! There are these little things zooming past. Zoom, zoom, zoom, and then a squid comes flying in! And then when you turn the lights off, there's all this luminescence. And afterward you realize there's so much that you don't know."

Such as the health of this ecosystem. It's not easy to assess the status of the ocean's microscopic creatures, she says. "You can't ask a zooplankton, 'Hey, how are you doing?' That's where the time series is vital."

A time series is simply a slow accrual of data. "They aren't flashy, and they're expensive," Amy says. The most revealing time series studies take generations to do, and the sheer length of time required makes them vulnerable to political and bureaucratic predation. Impatience can kill a time series project before its value becomes apparent, she says. But if you do them correctly, you can more accurately understand what happened in the past by factoring out seasonal fluctuations. "Then you can use this information to make predictions about the future."

Which, when it comes to the open ocean and its plankton, requires a deep breath.

---

It's helpful to know a bit more about phytoplankton, the forests of the sea. Coccolithophores and diatoms are important sun-loving forms of phytoplankton. You can see blooms of coccolithophores from space, milky swirls as large as California.

Under a microscope, you'll see they make fantastic shells. Shaped like flying saucers, they're made of calcium carbonate, the same thing as chalk. The White Cliffs of Dover are old deposits of coccolithophores.

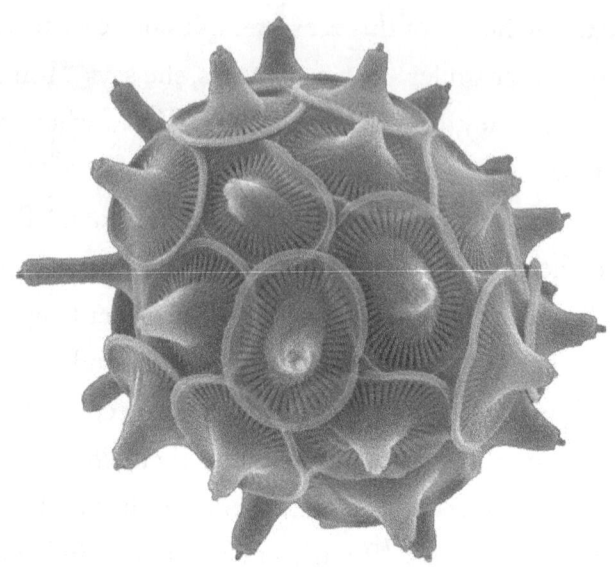

Microscopic view of a coccolithophore.
Photo by NASA.

Diatoms build glass shells that look like fancy hat boxes. Their shells are even smaller, so small that thirty can fit across a human hair. Medical examiners use diatoms in drowning cases: as victims inhale water, diatoms enter the bloodstream and make their way as far as the kidney and brain. If you find diatoms there, you know the person drowned instead of dying before entering the water. Diatom shells also are used in toothpaste, cat litter, dynamite, and nail polish.

Diatoms and coccolithophores don't just make useful and beautiful shells. They also incorporate huge amounts of carbon dioxide into their bodies. Then, larger zooplankton gobble them up. Or they die of old age, which in plankton time might be just a few weeks. Weighed down by their elegant shells, they fall toward the ocean floor, joining a chorus of other particles, including carbon-infused feces of zooplankton. Scientists call this descending material marine snow.

Marine snow piles up unseen on the ocean floor. And over millions of years, geological and other forces compress it and turn it into oil. This ongoing plankton bloom-and-bust represents a massive carbon pump to the ocean floor that scientists have only begun to fully understand. Every year, according to recent studies, phytoplankton incorporate fifty billion tons of carbon into their cells. That's roughly the same as all the forests, bushes, and grasses combined. But the Bermuda time series and other research have shown major changes in this biochemical balance. We've unlocked vast amounts of carbon dioxide by burning coal, gas, and oil—carbon dioxide that took millions of years to sequester. And this has made the ocean more acidic—about 30 percent more acidic since the Industrial Revolution. And it's happening faster than any known change in the sea's chemistry during the past fifty million years.

That's bad news for some plankton. Amy Maas' research on sea butterflies has shown that even slight increases in acidity turn their shells from clear to opaque. Longer exposure, and the shells dissolve. Corals experience similar fates. Some researchers predict that because of ocean acidification, most of the world's coral reefs could be dead within half a century.

But there's another twist. To the surprise of scientists, coccolithophores, the chalk-making phytoplankton, are blooming at unprecedented rates in the North Atlantic—a ten-fold increase between 1965 and 2010. It's unclear whether this is good news or bad. On the bright side, such blooms are sucking up some of the carbon dioxide we release when we step on the gas pedal. On the other hand, it's a telling sign that we've pumped so much carbon dioxide into the air that our atmosphere and oceans are saturated. "Something strange is happening here, and it's happening much more quickly than we thought it should," one of the study's authors wrote.

Finally, we have the warming ocean, which could affect plankton in an indirect but deadly way. Many species do just fine in the sea's naturally changing temperatures. But they also depend on hidden underwater currents to stir up nutrients from cooler lower depths. This upwelling of nutrients is the equivalent of marine compost. But hot air hovering above the ocean creates a thick layer of warm water on the surface. And this turns that layer into a cap, preventing the upward movement of nutrients; with no mixing, phytoplankton starve.

You can outrun rising sea levels but with less oxygen, you won't run far without gasping for breath. Some scientists predict that because of global warming the ocean's temperature may rise eleven degrees Fahrenheit in the next seventy-five years. Under that scenario, many of the ocean's mixing currents may disappear, researchers at the University of Leicester in England reported last December. With less oxygen from the ocean's phytoplankton, the study's authors said, the planet will experience "mass mortality of animals and humans."

"Are we screwed? People ask me that a lot," Amy Maas says as the storm outside lets up. But she's an optimist; humanity has a choice about reducing carbon dioxide emissions, reducing the size of that heat-trapping cap. "We're intelligent organisms that can come together and make good decisions." In other words, we're not drifting like the plankton, subject to some external current.

Back in South Carolina, I join Dennis Allen and Paul Kenny on another trawl. They point their boat toward a place with hidden sandbars and oyster reefs called No Man's Friend. After nearly forty years working these waters, Dennis has answers to many questions he posed to his grandfather so long ago. But as will happen, answers only triggered more questions.

The ocean hides its secrets well: the up-and-down movements of zooplankton that create the deep scattering layer; the movements of fish from the estuaries to the open ocean where they spawn; then the huge unseen migration of planktonic fish larvae back to the estuaries, where they grow into snapper, croaker and shrimp.

"Remarkable creatures," he says. "I've been doing this for thirty-eight years and I still get excited by what the net brings up."

Which happens today. During a trawl, they find a catfish in their net.

A catfish in a salt marsh?

A surprise, but should it be?

The previous winter, stunned scientists recorded temperatures at the North Pole above freezing. Joggers in New York ran shirtless on Christmas Day. "I just returned from a conference in Portland," Dennis says, mentioning a moderator who summed up the meeting's findings: "he says, 'Some wacky things are going on in our estuaries.'"

Ecology is all about relationships and understanding connections. Pull one string of the food web, and another part might unravel. Pull lots of strings, and sometimes those strings get tangled in a wad, which happens today.

During their first trawl of the previous year, they didn't catch a single fish. Which is normal. Many fish are supposed to be in the open ocean now, spawning, laying eggs that will grow into planktonic larvae, the future tide-riders to the Carolina marshes. But on this trawl, their boat slows ever so slightly as the net behind it fills. They pull up the net, Allen straining, four decades of sampling taking its toll on his back.

The net is loaded with fish. Not just one school, either. Many schools of different fish.

"This is an event," Paul Kenny says, his voice suddenly firm. The fish aren't supposed to be here now, and certainly not in these numbers. "This has never happened before at any time of year—winter, spring and fall."

Not in thirty-five years.

Dennis nods.

Must be the warmth of the waters, the change in the salinity—many factors probably hemmed them into the coast, he says. So many strings of the food web being pulled at once.

Later, as I drive back to Charleston, my mind is filled with questions: what other mysteries are hiding in the waves?

I take a deep breath and let it go.

And I thank the plankton for my second breath.

Photo by Jared Bramblett

Bleaching in American Samoa in 2015. Photo by Ocean Image Bank.

# FADE TO WHITE

IN WORLD WAR II, THE SECRET DIVISION 6 SCIENTISTS WERE among the first to probe the mysteries of plankton. But as I learned about jellyfish and diatoms, I came across an even older war story, one that would lead to another underwater cosmos that was surprisingly close to home.

The story begins in 1914, when the world was as tossed as the sea. World War I had just begun, and Europe's armies were digging in. Off the Carolinas, a storm turned the Atlantic into green froth. Amid the waves, a wooden schooner limped toward Charleston.

*The Frederick W. Day* was 170 feet long, a four-masted monument to the end of the days of sail. From New York, the ship set off for Wilmington, North Carolina, its hull crammed with sacks of Portland cement.

But off the Outer Banks, the ship struck an unknown object. Seawater flowed in. Winds shoved the schooner away from Wilmington, and the captain made for Charleston, his crew engaged in its own trench warfare against rising waters. Then the pump gave out; the water hit the cement powder, and the ship went down like the stone it would become. The crew made it safely to Charleston

Harbor in a lifeboat. Then time passed, and for nearly a century, memories of the shipwreck's location faded.

But the sea's memory was better, and soon tropical corals found the hardened cement. The shipwreck evolved into a northern outpost of the Caribbean, a coral reef with barracudas, spiny urchins and colorful tropical fish you might see in an aquarium.

Then, this young reef fell ill. The corals turned ghostly white, like a sick person's limbs.

What happened wasn't a mystery, though. At least not to Phil Dustan, a marine biologist at the College of Charleston. He'd seen the same thing happen in so many other stunning places.

---

Coral reefs are the ocean's masterpieces, among the oldest and most complex ecosystems on Earth. The corals themselves are tiny cup-shaped animals called polyps. Polyps have a jelly-like consistency with a whitish limestone skeleton that clings to hard surfaces, such as the *Frederick W. Day*'s cement bags. A polyp has venomous tentacles that pop out at night. With these tentacles, polyps snare zooplankton and other prey. Polyps also have a deal with microscopic algae that live inside them. The polyp shelters the algae, and the algae make oxygen and sugars for the polyp. Alone, the polyp is translucent. It's the algae that give corals their wild hues.

Married like this, polyps and algae build elaborate structures: the staghorn coral's antlers, the cerebellum-shaped balls of brain corals. Gold, green, and brown, most corals grow slowly, an eighth of an inch to four inches a year. But over time, this layering adds up. Australia's Great Barrier Reef is the largest living structure on the planet. Coral reefs off Hawaii are more than eighty feet tall.

Corals, however, are picky about real estate, preferring clear water and stable temperatures. If seas get too warm, algae produce too much oxygen. This triggers a rift in their fragile symbiosis. Sensing a threat, the polyp expels its algae like a betrayed lover.

But without its partner, the polyp fades to its translucent self, and its white reflective limestone skeleton turns bright white. This is what's known as bleaching. And across the world's warming seas, this phenomenon has destroyed one reef after another. When bleached for long periods, corals shatter and fall to the sea floor.

Yet, as with cancer patients, reefs can survive bleaching episodes; when conditions return to normal in time, the symbiosis begins anew. Then the colors come back—the greens and yellows and pinks that seduced so many scientists and divers, including Phil Dustan, who told me his quest to understand corals began when he nearly blew his head off.

Phil Dustan uses a plumb bob to make measurements in Bali in 2015.
Photo by Phil Dustan.

We're in his office at the College of Charleston, which has a stunning view of the harbor. The room is cluttered with books and photos, including one of Jacques Cousteau. Phil is a wiry man with short graying hair and eyes the color of the blue water he dives in. I ask how he became a scientist. He says that as a child, he had two loves: the ocean on Long Island and fixing engines. When not surfing or in school, he worked at a marine engine junkyard. One day, the owner asked him to fill up a truck tire.

"I knew a lot about engines but not much about tires. I filled the tire when it suddenly exploded."

The tire and wheel hit his forehead, peeling his scalp back two inches. Another section knocked him backward into a truck. For a moment, he felt himself leave his body. He saw a white light around everything. That night, doctors told his mother he probably would die by morning.

But the double vision in his blue eyes moved back to single, his memory improved, and in some ways, he was a better person for it all. Before, he was a middling college student. After, he was an A student. "It was an internal thing, that life is real. I was a little more serious." His senior year in college, he took money from a workers' compensation settlement to buy a science course in the Bahamas during Spring Break.

There, while snorkeling, he had his first look at an offshore coral reef.

"I was stunned. The water was crystal clear"—so different from the green juice off Long Island he'd surfed in. And the blues?

He'd learn later why the water was so vibrant: when light hits tropical reefs, it bounces back to the surface, hitting the surface's underside like a rubber ball against a ceiling. This bounce amplifies the light and anything it strikes so it feels as if you're suspended in blue. In this light, a healthy reef almost glows. And when you swim under the boat, light beams fall like waterfalls around the shadow. This shimmering light exposes a whole new world: translucent plankton that cast great webs; ctenophores with rainbow-colored filaments. The invisible is suddenly visible.

But as he explored the Bahamian reef that spring break, he only knew that it was wonderful. Suddenly, he spotted a change in the light. In the sand below, giant wings appeared. And then a black-and-white stingray rose and swam in a grand arc, and the elegance of all this sealed something inside his mind: love for the reefs and a driving need to understand the mechanics behind their beauty.

As a graduate student in the early 1970s, he would study reefs in Jamaica's Discovery Bay, reefs that left him grinning as he climbed back on board after dives. And from Jamaica, he landed a job with the Smithsonian Institution on Florida's Key Largo.

"The reefs off the Keys then were amazing, the biggest and most diverse in the United States, with a few that even looked like the ones I knew in Jamaica."

Back then, he imagined coral reefs to be vast and abundant—ecosystems with the kind of permanence you might expect from something built of limestone.

But the more he studied the reefs, the more he realized how fragile they were.

"If you want proof, you don't have to go far," Phil tells me. "Just go to the Florida Keys."

---

A few weeks after my visit with him, I'm in a boat off Key Largo. Towering thunderclouds billow on the horizon as mangroves give way to open water. The water's reflection flashes like the blue and silver scales of a fish. It's mid-July, and the water is eighty-seven degrees, unusually warm and a tipping point for corals.

"I can guarantee you'll see some bleaching," says Jason Vogan, a dive boat captain with Quiescence Diving Services. He steers toward one of the most popular and devastated reefs in the Keys.

The Florida reef system traces an arc south of Miami to the Dry Tortugas. It's the third largest barrier reef in the world, and it's the engine of a $6.6 billion tourism economy that supports 70,000 jobs.

Reefs like this also are important nurseries for the seafood we eat. Ocean bottoms are mostly sand, but coral reefs are the oases. Reefs cover less than 2 percent of the seabed, but one-quarter of the ocean's species depend on reefs for food and shelter.

Four miles off the coast, the blue turns darker in splotches, signaling a shallow reef. Jason slows the boat near a mooring buoy. Joining four other snorkelers, I jump into the blue world below.

Under the surface, brown-and-blue sea fans wave back and forth like Hawaiian dancers.

A shimmering school of minnows pours through a tunnel of coral in a twisting swirl. A five-foot gray shark suddenly cruises past. I start looking behind me, then turn back toward the carnival below, different corals: brain and elkhorn. They're easy to pick out because some have turned white from bleaching. And beside them in a sandy area are remnants of what was: staghorn coral antlers piled up like old bones.

And yet, many colors remain: yellow-and-white zebra fish move to and fro, as well as a school of blue angel fish. Even the bleached corals are beautiful. Some are as white as sugar.

Amid the colors, my mind drifts back and forth. This place is beautiful even with the bleaching coral. What's all the fuss about?

Later, I mention this to Phil Dustan.

"Yeah," he says with a nod. "You don't know what it was like before."

---

Phil arrived in Key Largo in the mid-1970s, when the reefs had a glow that reminded him of the flush energy of a healthy pregnant woman. Water around Carysfort Reef, one of the largest in the Keys, was clear. Visibility was about eighty feet.

He looked different back then as well. His hair was long, and he sometimes wore mismatched socks, but he had a growing reputation for his knowledge about reefs, so much that in 1974 Jacques Cousteau had asked him to join an expedition in Belize. During one wine-soaked lunch on Calypso, Cousteau lamented how human activity had killed reefs in the Red Sea. Phil thought at the time: "Cousteau's out of his mind. Reefs were too robust and stable, right?"

But as Phil studied the reefs off the Florida Keys, he realized that Cousteau was right. Reefs were like a body fighting off multiple diseases. Developers were carving canals through ancient corals, sending plumes of sediment offshore; septic tanks leaked raw sewage; commercial and recreational fishermen scooped up vast numbers of fish. Sooner or later, the reefs' immunity would suffer. But how could he track its health?

He and his colleagues came up with a plan. They staked off rectangular areas in the reef with stainless steel rods. They called these staked areas "transects." By analyzing what happened inside the transects, he could estimate a reef's "coral cover." It was like measuring the density of a forest's tree canopy. In the mid-1970s, Carysfort was a dense undersea forest with about 60 percent covered in corals. It seemed like a stable system, except for the kidnappings.

---

In 1983, a mysterious disease attacked black sea urchins across the Caribbean. In short order, as much as 97 percent of the urchins died. It was a huge blow to coral reefs because urchins munched on algae. Think of an aquarium without a filter. Without the urchins, algae soon smothered corals. The antlers of the elkhorn and staghorn corals shattered and fell to the sea floor.

At the same time, worldwide carbon dioxide levels continued to rise, trapping heat in the air and sea. The ocean grew warmer by about a degree through the 1980s and 1990s. And as the ocean absorbed more carbon dioxide, its pH changed, and seawater became more acidic. Some reefs began to dissolve.

In 1981, Phil packed his bags to take a position at the College of Charleston, a new base to continue his studies in the Keys. In 1983, he visited his transect sites.

"It looked like a bomb had gone off."

He measured it again in 2000. By then, the coral cover at Carysfort Reef had declined by 92 percent.

This destruction filled Phil and other longtime divers with sadness and rage. It was an environmental catastrophe as serious as the burning of Amazonian rainforests, but one hidden under the waves. Phil likened it to a crime spree without any police to bring the crooks to justice. "Think of the reefs as a healthy town. Then, suddenly, someone kidnaps a doctor, then a priest, then a police officer, its garbage collectors. After a while, the fabric of the town frays. Especially if nobody does anything about it."

He published his dire findings, which he said didn't square with Florida's efforts to market the Keys as a pristine diving destination. With little explanation, he lost a major source of research funding. "That's how it's done. You're marginalized." But other researchers uncovered how acidification was eating away at Florida's coral reefs faster than the corals could lay down new layers of limestone. "Lots of scientists think that ocean acidification is not going to be a problem until 2050 or 2060," says Chris Langdon, a marine biology professor at the University of Miami. "This is happening now."

In 2013, Phil Dustan went back to Jamaica's Discovery Bay. Its once-magnificent reef was a pile of rubble.

In 2014, he was back in Key Largo. "What's left of the reefs looked like a war zone."

Which is why he felt so strongly about the *Frederick W. Day*.

---

The wreck of the *Frederick W. Day* in 1914 was just eleven miles from the mouth of Charleston Harbor. Even so, nautical charts had it in the wrong spot. So, no one really knew where it was. As the decades passed, ocean and time consumed the ship's wood, and Oculina corals took hold on the cargo of cement. An undersea forest bloomed. It had puffer fish, amberjack, grouper, damsel fish, and manta rays. All just a relatively short boat ride from Charleston's docks.

Then, one day in 1982, a shrimper snagged his net on the reef. He called a local diver, Eddie Phillips, who called Tom Robinson, owner of Charleston Scuba. Robinson dove on it and was blown away.

During later dives, Robinson saw sand sharks, some larger than ten feet, green moray eels, "tons of urchins," tropical sponges, and a barracuda that was about six feet long. "It looked like a Dr. Seuss picture or the Beatles' *Octopus's Garden*." He spotted thumping shrimp, a species that can strike with enough force to break aquarium glass. He swam with ctenophores, which floated by like iridescent Christmas ornaments.

But word was out now about the wreck's true location, and the kidnappings began. Divers with spear guns took out the larger fish, including the sand sharks and grouper. Robinson, who was against spearfishing, also watched as fishermen dangled hooks over the reef. Beer cans landed on the sea floor. Fishing lines wrapped around the coral. The *Freddy Day*, as it's now known, lost some of its equilibrium.

One day, a student brought Phil Dustan a video she took while diving on the shipwreck.

There it was, what he'd seen in the Florida Keys, in Jamaica, and now here, bleaching coral.

"We're doing science wrong," Phil says as we talk about the Freddy Day. Scientists are too fearful of seeming like activists, but when the facts show something catastrophic is happening, scientists must speak out, he says. They're the ones with the facts. He tells me about his recent trip to Bali with a group of students. "I had tears in my eyes when I came back to the boat." Warm water had triggered a major bleaching event. In just a year, bleaching and other forces destroyed one of the most beautiful reefs he'd ever seen. "If we don't step up as humans, it's game over. What makes me angry is knowing what's wrong and doing nothing to fix it. That's what I struggle with. But I tell my students that there are things we can do."

"What do you tell your students?" I ask.

"Focus on solutions," he says. "Do small things that add up. Vote for politicians who treat climate change seriously. Vote with your pocketbooks by buying more environmentally friendly products. Let's grow sea urchins. Or grow corals in nurseries and transplant them." He tells students that they should push politicians to create marine sanctuaries, as Tom Robinson would love to do with the *Freddy Day*. Do these things—reduce these smaller stresses—so reefs can handle the larger ones, such as a warming ocean and an increase in the sea's acidity. "Like an immune system, the healthier it is, the better it is at warding off diseases," he tells me. "So, there's hope. But hope has to be tied to action."

Without action, he says, more reefs will lose their fragile symbiosis, and some of the world's colors will fade forever.

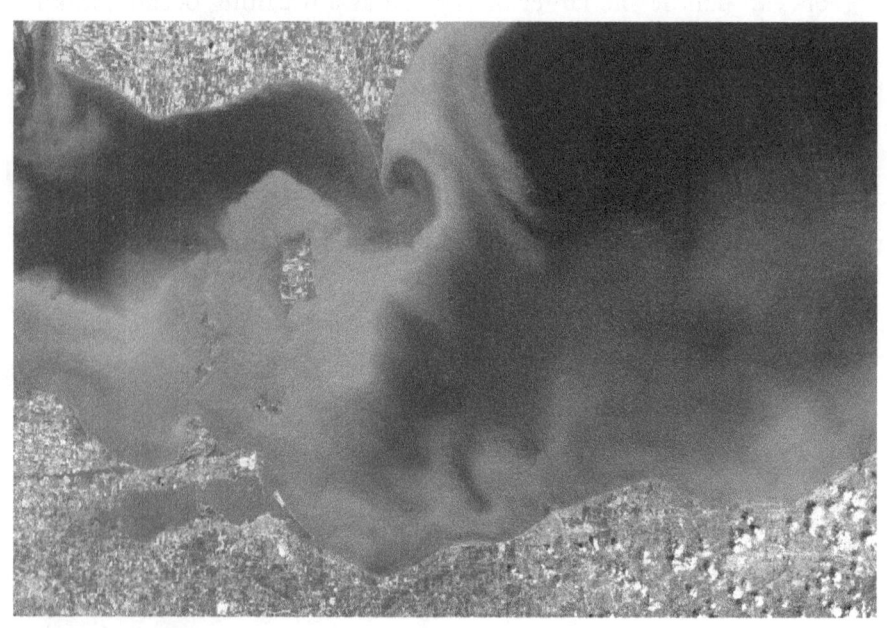
Algae bloom on Lake Erie in 2015. Photo by NASA Earth Observatory.

# SCUM

NOT LONG AFTER I SPOKE WITH PHIL DUSTAN, MY PHONE RANG. A man with a booming voice said he was a retired general and had a tip that affected the nation's security. "Tell me more," I said, and he launched into a long discussion about algae.

Seas, lakes, and ponds across the world were "going green" at an alarming rate, he explained. Hyperactive algae are the culprits—scum that can generate toxins as deadly as cobra venom. Climate change had injected massive amounts of heat into the ocean and lakes, creating giant Petri dishes for algae. Areas prone to algae blooms were seeing longer "bloom seasons." In the United States, algae blooms had smothered manatees in Florida and wiped out sea otters in California. One algae bloom in Florida turned thirty-three square miles of Lake Okeechobee the color of guacamole.

But what did that have to do with the nation's defense?

Hidden in each bloom was a potential mystery, the general answered. Algae are chemical warriors, and when they bloom, they sometimes create new toxins, like an army that develops a new weapon. In biological circles, these toxins were "new to science," substances yet to be discovered, measured, and named. Foreign countries would love to get their hands on these deadly substances, the general said, adding that only a few scientists in the world had the detective skills and technology to discover these hidden toxins. And one of those rare scum sleuths happened to work in Charleston.

His name was Peter Moeller, a scientist with the National Oceanic and Atmospheric Administration based at the federal Hollings Marine Lab on James Island. The general heard that the federal government was thinking about shutting down Peter's lab, and he feared that doing so would weaken the nation's ability to identify potential bioterrorism weapons. I called Peter that afternoon.

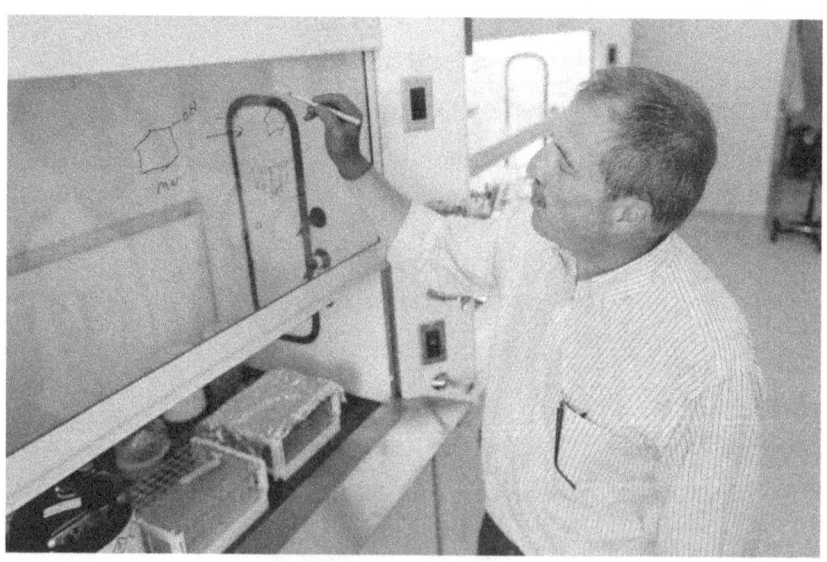

Peter Moeller in his lab at the Hollings Marine Center.
Photo by Chris Hanclosky of *The Post and Courier*.

Peter Moeller had a thick mustache and wore a white shirt with glasses peeking from the pocket. He spoke quickly and tended to make quick conversational turns, like a driver rushing to parallel park. He's from Minnesota, the Land of Ten Thousand Lakes, a number of which now see regular algae blooms. Yet, growing up, Peter told me he was more interested in dirt than algae, specifically, his father's dirt, and why his old man put it in the mail.

His father was a farmer. That package of soil? Sent to be analyzed. "Then he told me about the chemistry. He told me about bacteria and organisms, and then I saw the results out in the fields." Some crops did well in certain soil conditions. Others wilted. Peter also watched a lot of Jacques Cousteau on TV, so he was interested in the ocean as much as dirt. When a sixth-grade teacher asked what he planned to be when he grew up, he put it all together in his answer—marine natural products chemist. He didn't really know what such a chemist did, but something about it sounded right—how the ocean hid secrets to medicines people might someday need. So, from there, he became what his mind conjured as a child: a chemistry student at Mankato State; master's and PhD candidate at the University of California, San Diego; researcher in British Columbia, specializing in novel toxins.

"How did you end up in Charleston," I asked.

A scientist at NOAA invited him in 1992 to visit Charleston and consider starting a lab at a new state and federal lab cluster at Fort Johnson on James Island. The scientist asked how much a new lab might cost. Forty thousand? Sixty thousand? "I said, 'Probably four to six million dollars.' He looked at me kind of funny and didn't say much. So, I went back thinking that was the end of that."

Unknown to Peter, South Carolina's senator Ernest Hollings wanted a state-of-the-art science community in Charleston, and he had a well-deserved reputation for bringing federal jobs to the area. Congress eventually appropriated nearly fifty million between 1994 and 2005 for new labs at Fort Johnson, including one to identify new toxins—Peter's lab.

He and his colleagues soon acquired giant machines packed with superconducting magnets—room-size gizmos that break down chemicals to the atomic level. Construction crews drove pilings deep into the Lowcountry marl—deep enough so vibrations from passing trucks and other machines didn't skew results. "I had the toys I needed."

He and his colleagues began their search for things no one had ever seen before.

Like the mystery toxins in the plastic bag he pulled from a mini-fridge during my first visit.

The Ziploc bag contained dark green muck—freeze-dried algae paste. Peter wore a blue protective glove. He held the bag up to the light. Like an illegal drug, its contents were dangerous and valuable.

Pharmaceutical researchers often need toxins to determine whether their drugs are effective. But toxins are difficult to isolate and purify, and this makes them expensive. "If we pulled out all the toxins from just this bag, we probably could get four million dollars for those samples," he said. "But if I did that, there would be a lot more security people around here."

"Why?" I asked.

He smiled.

Something in that bag had killed before.

Fish, to be specific, found belly up in Lake Michigan in 2016, near a water intake. And this particular crime scene was full of green slime. The origins of this slime spree had likely begun on land. Storm runoff carries manure, fertilizers, and pollutants into a body of water. In the water, algal communities feast as if presented a free buffet, gorging themselves and reproducing at astonishing rates. Lakes can turn green in a matter of days. In the summer of

2014, four hundred thousand residents in Toledo, Ohio, woke up to warnings that their drinking water wasn't safe. The perpetrator? An enormous algae cloud on the city's water intake on Lake Erie.

When blooms die, they often leave behind cysts. These cysts remain dormant until waters warm again. Then, these microbes revive like zombies. "Once a lake goes green, it often becomes a chronic problem," Peter said. Thus, Lake Erie became the poster child of algae blooms. In 2015, another bloom appeared, creating a mat of scum about the size of New York City. It happened again in 2016 and 2017.

When Peter began investigating the Lake Michigan fish kill in 2016, the leading suspect was a blue-green algae that makes a toxin called microcystin.

Microcystin is common but nothing to fool around with. If you're exposed to enough microcystin, your lips and mouth can turn numb within half an hour. That numbness can spread to your face, neck and extremities. Then the poison hits your liver, and your brain. Lower dosages can trigger nausea, vomiting, headaches, and other flu-like symptoms. In the mid-1990s, fifty people in Brazil died after undergoing dialysis with untreated water laced with microcystin. Dogs, cattle, and fish are especially vulnerable. Hundreds of dogs have died because of their tendency to drink water while they swim, researchers with the Centers for Disease Control found. But these algae-created toxins may do more than poison fish and dogs; they may trigger diseases.

Researchers recently found higher rates of liver disease and cancer in areas near blooms. Investigators in France, Florida, and New England also have linked clusters of amyotrophic lateral sclerosis (ALS, or Lou Gehrig's disease) to algal blooms that spawned a toxin called BMAA. In one 2014 study, scientists found that living within eighteen miles of algae-prone lakes raised the odds of living in an ALS hotspot by 167 percent. Scientists have long sought to identify agents that cause ALS, Alzheimer's and Parkinson's. And in what some call a key piece of that "toxic puzzle," researchers have reported that monkeys fed food laced with BMAA developed Alzheimer's-like neurodegeneration. As climate change increases the chances of algal blooms, this issue has only become more pressing.

These and other studies also rest on an assumption: a known toxin is causing harm.

But what about the unknowns?

When officials test blooms, many often blame microcystin and other known toxins—and stop there, Peter told me.

"I call it science by convenience."

The real culprit may still be hidden in the scum. Peter estimated that as many as a quarter of all toxic events are wrongly attributed to microcystin and similar known toxins.

That's where his lab comes in. His job is to identify toxins that have done harm—compounds that no one has seen before.

And, in the case of the Lake Michigan fish kill, the mystery perps were still in that bag of freeze-dried muck.

---

From the Lowcountry's live oaks to California's kelp beds, all of Earth's life forms trace their origins to blue-green algae. It's technically not algae but rather a form of bacteria known as cyanobacteria. Swayed by the coloring, early scientists misclassified it, but the name stuck.

When it blooms, cyanobacteria sometimes looks as if someone emptied tankers of green paint into the water. With its vibrant green tint, it has its own beauty, even as it hides potential poisons. No matter the label, cyanobacteria is among the planet's first organisms to convert carbon dioxide into oxygen through photosynthesis. Under the right conditions of heat and nutrients, these microbes rapidly multiply. And when this happens, they go to war.

In their battle for survival, cyanobacteria, algae, fungi, viruses and other microbes invent new compounds to protect themselves and dispatch enemies. The more potent the better, which is why these microbes are so adept at making small amounts of highly toxic stuff.

Peter pointed again to the bag of freeze-dried mush from Lake Michigan. "There are literally billions of microorganisms in this bag right now, and there are probably at least three-thousand toxins in there, most of which are new to science."

Finding these novel toxins involves a lot of death. Using beakers, special evaporators and solvents, he and his colleagues separate algae paste into different classes of chemical compounds. Then they test them against cancer cells to see which compounds kill. "We're like a Stephen King lab. We like it when things die."

It's a process of elimination, like dividing groups into gangs—the Bloods and the Crips—and then doing tests to prove a suspect is a Crip, not a Blood. He and his colleagues do this "partitioning" over and over, narrowing the field. The process can take years.

Peter walked to another room. A window framed a view of the marsh toward Fort Sumter. The lab's mass spectrometers sat high on a counter. Mass spectrometers bombard materials with electrons to shatter them. Each chemical compound has its own unique way of breaking, a pattern that can be used like a fingerprint. "It's as if you're looking at 100 people, and you find only one who weighs exactly 110 pounds. You can find that person again using that weight." Over the past year, he and a graduate student had narrowed the Lake Michigan fish kill to two potential toxins.

Two very bad guys, he added. Nearly as poisonous as botulism and as much as a thousand times more toxic than microcystin, one of the usual suspects in fish kills. But they still had one more set of experiments to do that afternoon—ones that would reveal the two toxins' atomic structures—prove whether they were, in fact, new to science.

Over the years, Peter and his lab's partners have identified one new toxin after another, including poisons that could save lives. In 2002, a microbiologist named Paul Zimba learned about a fish kill near North Carolina's Albemarle Sound. Zimba gathered scum samples, grew algae cultures in his lab and sent them to Peter's lab.

Peter and his colleagues identified a previously undiscovered toxin. Zimba, now director of the Center for Coastal Studies at Texas A&M, tested it on cancer cells. To his surprise, he found the toxin killed a range of different cancer cells. Its unusual selectivity meant it could be possibly used in cancer treatments, especially colorectal cancers.

Pharmaceutical researchers aren't the only ones who can harness algae's creative chemistry.

Spy agencies have had a long history of harnessing algae's toxicity. In 1960, Francis Gary Powers, the U-2 spy plane pilot, carried a tiny injector when the Soviet Union shot down his plane. Hidden inside a silver dollar, the injector contained saxitoxin. Microalgae in red tides make saxitoxin, and the toxin can be taken in by shellfish, creating a health hazard. Purified, it's one thousand times more toxic than sarin gas. Powers was given the injector in case he preferred death to torture.

Peter said that with minimal chemical know-how, terrorists could disperse algae-created toxins and cause widespread illnesses and deaths—use toxins as bioweapons.

That brought me back to what the general with the booming voice had told me, that budget cutters might shut down his lab.

I asked Peter about this, and he asked me to look around and tell me what I didn't see.

The answer was people. Spend a few minutes walking through the halls of the Hollings Marine Lab, and you quickly notice the emptiness. A decade ago, Peter had eighteen scientists and students working in the lab. Now, he was down to one graduate student. If officials shut down the lab even temporarily, millions of dollars in equipment would end up in the junkyard. Two of the lab's nuclear magnetic resonance machines contain superconducting magnets. If maintained properly, these magnets spin in a kind of perpetual motion. If maintenance stops, the spinning stops, and the machines would be ruined.

I left his lab thinking his work seemed even more important amid a rapidly warming planet. Harmful algae thrive in warm water, and one study predicted the number of algae blooms would triple in the next thirty years. This would have real economic and social impacts. In the United States, harmful algae blooms cost communities an average of $50 million a year. A single bloom on Lake Erie in 2014 created a $65 million economic hit. But in many ways, we didn't have a good handle on the scope of the problem. I was surprised to find that the federal government didn't keep a national database of outbreaks. So, I began counting them myself, scraping data from state websites and local newspapers.

I found a case in Southern California where a toddler ended up in the emergency room after swimming in an algae-fouled lake.

I found other cases in California where blooms likely killed four dogs. An outbreak in Oregon killed thirty-two head of cattle, and dozens of sea lions in Northern California washed up dead with signs of algae-related poisons. I stopped counting when my algae bloom total reached more than five hundred in forty-eight states.

---

A week later, Peter invited me back to his lab to watch an experiment that might finally pinpoint whether a new toxin was responsible for that Lake Michigan fish kill.

He ushered me into the room the size of a missile silo. It contained a nuclear magnetic resonance machine. I sat in a control area. Signs outside warned about getting too close. Its magnets were so powerful that standing within twenty feet could erase data on credit cards or interfere with heart pacemakers.

While the mass spectrometers can expose a suspected toxin's molecular weight, the nuclear magnetic resonance machine, "gives us the picture of the guy, the fingerprints, everything. You have the culprit."

He transferred the contents of the vial into a glass straw—drops that represented six weeks of purification work. He handed it to Stuart Parnham, a British physical chemist. "You're not giving me much to work with here," he joked.

Stuart climbed a ladder to the top of the machine and placed the vial on a cradle. It bobbed on a cushion of nitrogen gas. Back at his computer, he pressed buttons. Like an elevator going down, the vial disappeared. The machine's magnets spun.

"Let's do some experiments," Stuart said, as Peter left the room to check on another experiment.

The screen filled with data, a mathematical portrait of a killer compound.

The one that had been hiding in Lake Michigan.

"You're seeing it first," Stuart said.

New to science.

Norman Levine, a College of Charleston geologist, dips his hand in tidal water. Photo by Wade Spees of *The Post and Courier*.

# LOWCOUNTRY ON THE EDGE

AT TWILIGHT, I MEET NORM LEVINE, A GEOLOGIST AT THE College of Charleston, and we step right into it—salty water oozing out from beneath the pavement.

"Very impressive," he says of the king tide. The moon is close to Earth, and its power pushed the Atlantic deeper into our twisting tidal creeks. But it's a foot above normal king tides, so water bursts through cracks near the very accurately named Flood Street in downtown Charleston's Westside neighborhood.

Norm watches the saltwater pool and shoot across the street like a rapid. A nurse in blue scrubs tiptoes through the current to reach her car. His voice rises because king tides are a taste of our future, one flavored with salt; he knows that climate change is reshaping the world's low places, and the Lowcountry also is well-named. "This isn't even a super king tide," he says. "Sea rise is no longer a probability, yo. It's happening in real time."

Norm has a round physique and a dark goatee framed by wire-rimmed glasses. Colleagues jokingly call him the "Voice of Doom," but his voice is nothing like Darth Vader's. He speaks quickly and with enthusiasm. His accent has hints of New York and the South, and he has a disarming way of tossing the word "yo" into conversations.

The doom part comes from his title as director of the college's Lowcountry Hazards Center. The center has seismographs to measure earthquakes, weather gauges to document hurricanes, and computers packed with terabytes of mapping data. He and his students have used some of these maps to make startling predictions about the Lowcountry's edges as the atmosphere and oceans warm.

From previous conversations, I knew Norm had long been fascinated by mega forces. He grew up in an area of Long Island smoothed by glaciers. He was captivated by how slabs of ice from what's now Upstate New York could transport gneiss, schist, and other glittering rocks to yards around his house. In middle school, teachers gave him the key to the school's rock collection. In college, he stared at images of Earth snapped by Apollo spacecraft, his mind transfixed by the blues, whites, browns—water, vapor, land. So, he became a geologist, learning quickly that the field isn't just about rocks. It's also about water and movement and time. He would learn, for instance, that the area around Charleston is sinking. Slowly, but surely, North America's main tectonic plate is tilting like a giant see-saw with the pivot point roughly on the Canadian border. Loosely speaking, the Canadian side is going up, and the South Carolina side down—down just a few inches a century, not such a big deal. Unless the land already is low and next to a rising ocean.

Back to water instead of rocks. Imagine a big pitcher of ice cubes with the water filled to the brim. Water won't flow over the edge as the cubes melt, even if it's a hot day. But add more cubes, and you'll have water pouring onto the table. That's what's happening in Greenland and Antarctica. All told, Greenland has the equivalent of twenty-three feet of sea rise in its glaciers and ice. Antarctica holds even more ice—two hundred feet of potential sea rise. Cue the Voice of Doom: "we have more ice cubes about to drop into the cup."

A Florida-sized ice cube on the western side of Antarctica is called the Thwaites Glacier. If it drops, seas rise two feet higher. The most recent studies say it's losing fifty billion tons of ice a year.

The impact is as real as the saltwater pooling by our feet, and these impacts are accelerating. In the past, scientists thought sea levels rose in a straight line. They've been measuring the sea level in Charleston Harbor continuously since 1921, and for most of the last century, sea rise roughly followed that straight line.

But in the mid-2000s, scientists realized that sea rise wasn't following a linear trajectory anymore. It was a sharp upward curve.

In Charleston, the pattern was unmistakable.

From 1990 to 2000, the sea level rose 1.4 inches.

From 2000 to 2010, it added 2 inches.

From 2010 to 2021, 2.7 inches.

Follow this upward curve into the future, seas would rise an additional 3.2 inches by 2030. And faster as more ice cubes fall.

"We know that sea rise is already accelerating, yo, and it's happening more sharply than they predicted ten years ago," Norm says. "And we're not talking in terms of one hundred years anymore." He watches more hospital employees wade through the Hagood and Fishburne Street intersection. "We're talking about big changes in decades."

"Like what?" I ask, and he tells me about a study he and his colleagues did in 2015. They analyzed satellite maps capable of discerning differences in a few inches of elevation. These maps quickly revealed that the Lowcountry is indeed flat but not evenly so. It's full of subtle rises and dips, more like an old plaster wall instead of the more uniform flatness of modern wallboard, or Florida.

They also discovered that just one foot of sea rise would flood 204,000 acres of marsh and 64,000 acres of land, about a quarter of Charleston County. Nearly one thousand homes, offices and other buildings would see regular flooding. Moving up, three feet of sea rise will inundate nearly 9,400 buildings. Six feet? More than 34,300 structures end up in the drink.

---

We leave Flood Street for another changing edge. We cross King Street, peninsular Charleston's slightly elevated spine. We drive into the gentrifying and slowly flooding neighborhood on the East Side. Water curls around a church, and floodlights from a nearby field reflect off the pooling water. Across the street, notes from a blues guitar pour through a wrought-iron gate made by Charleston's famous blacksmith, Philip Simmons. Longtime resident Santel Powell has an old John Lee Hooker album blasting into the night. We shout a hello, and Powell and the professor begin chatting about his house. It was built in 1882 on a platform of cypress and pyramid-shaped brick pilings, framing made with wooden pegs. The Voice of Doom gets excited, because the house was designed to take the brunt of both earthquakes and hurricanes. And it's also built on one of the Lowcountry's gentle rises, so the evening's floods don't reach Powell's property. We make small talk before we drift into a conversation about sea rise. Powell says he's in his sixties. The effects of sea rise feel far away, he says, beyond his life's horizon. "I'll probably have a heart attack before it gets to my house," he jokes. "Besides, what are you going to do? Pick the city up?"

"Yo, that's exactly the question," Norm says.

---

No place is better suited and more fun to answer that question than New Orleans. New Orleans is surrounded by water, and half of its land is below sea level. Its soil is as spongy as the Lowcountry's. This is a problem. As New Orleans expanded and paved over that soil, the sponge flattened. Wrung out, the city is sinking under the weight of its history. At the same time, rising seas and intensifying storms have pushed the city's levees and 120 drainage pumps to the brink and sometimes over it.

All this creates an urban cocktail of impending doom and end-time revelry. Raised tombs in cemeteries remind you that the ground is too low to bury someone in perpetuity. Roller-coaster streets and water main breaks remind you of the land's subsidence. From the raucous French Quarter, you see ships on the Mississippi moving eerily above you. Straddling a levee between denial and action, New Orleans offers important lessons for other low places, including Charleston.

One afternoon in December, Julia Kumari Drapkin shows off some of New Orleans' new efforts to adapt. Drapkin is founder of I See Change, a group that studies climate change's effects on cities. For much of New Orleans' history, residents saw water as an enemy force, she tells me. The city survived by adding defenses—building levees and pumping water from catch basins into canals and lakes Pontchartrain and Borgne. This system lifts as much as four hundred thousand gallons a second. Pumps run even when the weather is dry, collecting runoff in underground pipes and tunnels. Sometimes the pumps don't work properly. And pumps or not, rain bombs are simply too powerful to pump your way out of flooding.

A summer drencher can create thigh-deep floods across the city. "The system would need to be six times bigger to handle an event like that," she says.

So instead of focusing only on pumps, the city is building networks of water storage projects, perhaps enough to let more water seep back into that flattening sponge and slow the land's slow descent. One prominent project is the site of a former convent. Hurricane Katrina flooded the Sisters of St. Joseph community in 2005. A few years later, lightning struck, triggering a fire. The nuns decided enough was enough and donated their twenty-five acres to the city. Now, planners have begun developing the Mirabeau Water Garden, a park that will double as a holding pond during severe storms. The convent is gone now, leaving behind a sidewalk in the shape of a cross and a field of live oaks and diverse grasses. Plans call for gentle ditches that turn into temporary ponds during storms. When drier weather returns, they revert to fields and swales. Planners estimate the storm garden could remove 10 million gallons of stormwater from more than three thousand acres around it. So far, federal grants have paid $12.5 million for the park, about half its expected cost. Major construction began in 2023.

Another, smaller project is nearby and not far from a levee that failed after Hurricane Katrina. Many neighborhoods still have vacant lots, including two in the Gentilly Resilience District that became a small rain garden.

A city official tells me the park-like plot can capture about eighty-eight thousand gallons from nearby streets. It holds it for a short period in a donut-shaped pond until the pumps kick in. When you tie this into a network of other water collection plots, you buy the city's beleaguered pumps more time. The city also is promoting everything from rain barrels to planter boxes that catch stormwater.

It has a program that helps homeowners remove pavement so their yards soak up more rain. Across the city, I spot front lawns with mounds of dirt waiting to be spread – homeowners waging war against rising waters one inch of topsoil at a time.

But sometimes you need to move faster—and higher.

---

Cross the arrow-straight causeway from New Orleans, and you land in the small city of Mandeville. I meet Roderick Scott near the town's seawall and mention what I'd seen in New Orleans. He answers with a shrug. "It's a drop in the bucket of what needs to be done."

Scott launches into his story. He was a contractor in Iowa when severe floods hit parts of the Midwest in 2008. Sensing a business opportunity, he shifted his business toward rebuilding flood-prone historic structures. He eventually moved to New Orleans to help start a company that elevates buildings. That company ultimately raised more than fifteen hundred homes and businesses in Louisiana, he says. He's since become a sought-after flood mitigation specialist. He'd recently visited flood-weary Charleston to talk to preservationists about the dos and don'ts of raising old homes.

"When you're talking about New Orleans, Charleston, Miami Beach, tourists go to see the old buildings, right?" he says. "And if you get rid of the old buildings, what do you have? So, your options in these cities are limited. There are basically three forms of mitigation: elevation, relocation, and demolition." He falls squarely on the side of elevation.

Mandeville has about thirteen thousand people and an average elevation of six feet. Its historic district is on the edge of Lake Pontchartrain, which because it's open to the Gulf of Mexico, is at sea level. Like Charleston, this low-lying area floods often. When southwest winds blow hard, waves pile onto each other and crash over the wall. Same with hurricane storm surges. Mandeville doesn't pump water like New Orleans. Instead, as with some areas of Charleston, Mandeville's drainage system depends on gravity and time.

Scott points to the old Don'z bar across from the seawall and suggests we take a look. The structure was built in 1848 and once housed a Civil War hospital on the second floor. The bar is on the ground floor, and inside, a patron named Bobby Boettner says he's seen firsthand how floodwaters could rise to the level of the barstool. He says he knows this because he kept drinking while the water flowed in during a hurricane. "I stayed right here on this seat. I had a cooler floating next to me," he says with a proud grin.

Scott says ground-floor structures like Don'z are vulnerable to more than rising waters. "Climate change isn't the problem at the moment. It's flood insurance," he says. Homes that aren't raised high enough to meet federal flood insurance guidelines are seeing dramatic rate hikes. More increases are on the way as the National Flood Insurance Program tries to right its balance sheet. A small ground-floor business like Don'z could end up with an annual bill of $45,000 in a few years. "It's going to price people out of older buildings." He's already seen effects on the real estate market in Mandeville and other areas of Louisiana. Homes that are raised usually sell quickly and at escalating prices. "But the ones on the ground—you can't sell them, or they go minus the cost of elevating them. Homeowners here know this better than anyone, and they did something about it."

I ask why Mandeville acted so swiftly.

"Taxes," Scott says.

Residents in ground-level homes demanded that their taxes be lowered because flooding was ruining property resale values. City leaders listened and lowered their taxes. But lower taxes meant less revenue, and this had a ripple effect on everything from schools to city departments. Suddenly, the city realized that rising seas could drain the city's budget. The city loosened its regulations to raise homes, even in older neighborhoods. So far, tiny Mandeville has raised about four hundred structures in vulnerable flood zones with about one hundred to go.

"To be sure, we've had our share of growing pains," Scott says. Early on, Mandeville had few guidelines about raising homes, even historic ones. And this led to some aesthetic mistakes. Some homeowners raised their homes with metal pipes used in the oil industry, which makes them look as if they were built on playground equipment. Others raised homes on unfinished concrete blocks, exposing the insulation and plumbing, like a person without pants. Over time, Mandeville pushed for better designs. Newer approaches include lattice work and outdoor drapes. City officials also urged homeowners to create more attractive front stairways to make entrances more inviting to pedestrians. Another lesson was to hire architects and engineers early in the elevation process. Too many residents ended up with homes that lacked proper entrances or lacked natural connections to streets and sidewalks.

Scott estimates that three million homes across the nation are candidates for elevation, relocation or demolition because of rising seas and other factors related to climate change. Mayors and councils and planning officials that fail to protect their constituents' homes from flooding will lose their jobs, he predicts. "Mandeville, Charleston, and Miami—these cities don't have much time, a thirty-year-mortgage or two, to defend themselves against the sea. We need to fix them for our children."

As we walk around an old neighborhood, the elevated old homes look stately enough, but the street also has the feel of a barrier island. Something is lost when a home loses its connectivity to the ground. But what's the alternative? Having coffee on the porch in your waders?

---

Back in New Orleans, more than twenty-three thousand Earth and space scientists gather for the world's biggest nerdfest. It's the fall meeting of American Geophysical Union, a conference so large that the group only holds it in cities with massive convention centers. New Orleans qualifies, and this meeting is set in an exhibition hall that runs for a mile along the murky Mississippi and has the space of nineteen football fields. Over the course of a few days, scientists attend more than two thousand talks. The hall is packed with elevated bulletin boards lined in rows as long as city blocks. But instead of street names, signs at intersections say things like "Cryosphere" and "Tectonophysics." Attached to the boards are posters that distill attendees' discoveries. Scientists stand next to them and take questions. Imagine thousands of people itching to talk about the aha moments in their labs, and you begin to get a sense of the energy in that hall.

I wander down the canyons of posters in awe of this energy and the work behind it all. Groundbreaking science takes time, especially when it's about something as complex as the Earth's geology and atmosphere. If you're a scientist, you might spend years pursuing a hypothesis, years before a spark lights a promising path. Then you must fan the spark so it doesn't fizzle—test your discovery with experiments, computer modeling and reviews by your peers. Then other scientists must repeat what you did, because repetition is proof. Scientists are detectives at heart. They search for evidence about how the world really works. They debunk myths. To me, there's a form of justice in this process, which is why I like talking to scientists so much. Who doesn't like a good detective yarn?

But there's one problem, and I hear it as soon as I stop by one of the poster talks. I don't speak Sanskrit, or whatever language of their field they're speaking. I've found scientists have trouble simplifying their work, because they deal in nuances and caveats. I've often asked scientists to explain their work by pretending I'm in middle school, and when that didn't work, kindergarten. But bridging the gap between science and the public has never been more important, given the stakes of a rapidly warming planet. So I charge deeper into the scrum, heading for talks about hurricanes and jet streams.

I find several research groups who focused on Hurricane Harvey, the tempest in 2017 that dropped fifty inches of rain on Houston. One international team used computer models to estimate the likelihood that global warming made Harvey more catastrophic. They found that climate change increased rainfall by nearly 20 percent. Another study said a storm of Harvey's intensity would have been a once-in-nine-thousand-year event in a pre-industrial climate. A third linked the storm's punch to heat stored in the Gulf of Mexico. Had the heat content been lower, Houston would have been hit with less rain.

A pair of studies from the Massachusetts Institute of Technology identified changes in the jet stream—the river of air that undulates across the continent. A rapidly warming planet may slow the jet stream, causing deeper bends—and in Harvey's case—a storm that stalls and dumps more rain than it would otherwise.

Another group of federal scientists says the Arctic was "going through the most unprecedented transition in human history." Another team of scientists says that three extreme weather events in 2016—a massive blob of warm water off Alaska, heat waves in Asia, and high overall temperatures elsewhere—were linked to the burning of fossil fuels and other human sources.

It's grim news, which creates a bit of cognitive dissonance as I walk from one session to another. You can see the joy in the researchers faces as they talk about their discoveries. You can feel the vein of positivity embedded in their conversations: research matters, repetition of results matter, facts matter, working together on new projects matters. Because a more certain world is a more predictable world—and by extension—a safer one. But many of the studies I learn about carry a different message: the pace of climate change is accelerating, and this is making the world even more unpredictable.

So, by the end of the day, I'm a little dizzy from all this positive and negative energy. I'm a little relieved as I leave the exhibition hall. I follow packs of researchers carrying poster tubes. We cross a street to a lively bar that touts its hurricane drinks of rum and fruit juices. In the distance and above, I see a levee that protects the city from the Mississippi and the Gulf of Mexico—an ever-present reminder that this precarious city's future hinges on aging levees, sketchy pumps, and hope.

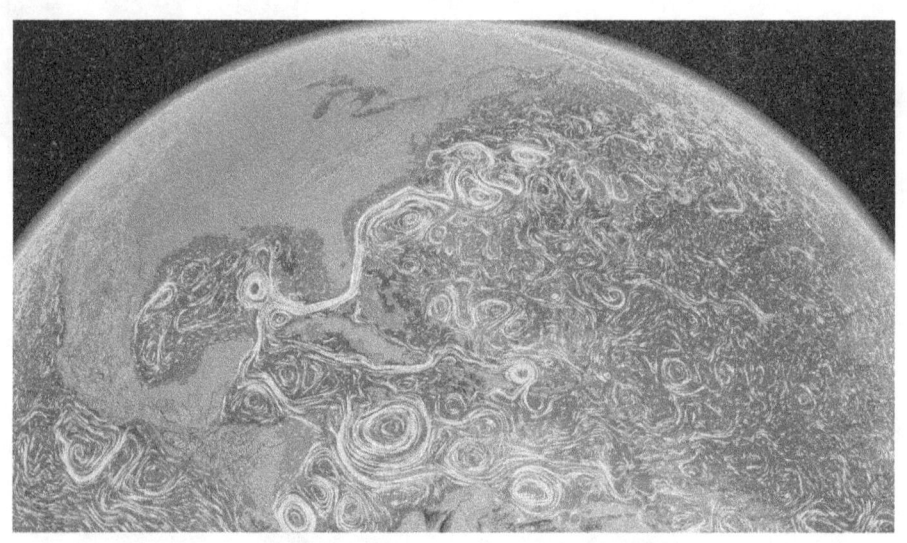

Rendering of the Gulf Stream's flow. Illustration by NASA.

# INTO THE GULF STREAM

DAWN BRIGHTENED THE SALT MARSH AS A NINETY-FOOT-LONG fishing boat left Little River, a small town about twenty-five miles up the coast from Myrtle Beach, SC. Aboard were about sixty people, including Jeter and Kathy Allred, also known as the Grouper King and Queen.

"I've been out in the Gulf Stream when it's so blue that it was almost purple, and you could see a hundred feet down," Jeter, the King, told me as we headed east. People napped as the swells grew larger. Then after about three hours, the ocean changed color, but not quite as dramatically as the Grouper King described. Instead of purple it was bluish green. Winds had blended the normally cobalt blue waters of the Gulf Stream with murkier inshore currents. Passengers began moving quickly, as if an alarm clock had sounded. As I studied the colors, they lined both sides of the boat, holding rods like rifles, silently waiting for the captain's okay. When he gave it, everyone dropped their hooks, as if in a race. Within minutes, anglers brought up crimson snapper and glistening grouper. The deck filled with fish and blood from black sea bass and grunts. Blood oozed off the deck back into the water. I didn't want to fish.

I'd booked a ride just to get into the Gulf Stream itself, a force so powerful that it lowers our sea level. But the guts were getting to me.

As I tried to steady my stomach, my mind drifted back to a story about Ben Franklin.

---

Before the American Revolution, Franklin was a Colonial postmaster. He'd noticed that British mail ships were slow. Much slower than other merchant ships. He mentioned this to his cousin, Timothy Folger, a ship captain who hunted whales off New England. Folger knew exactly why—a powerful ocean current. Folger even warned mail ships to stay clear. But, he told Franklin, the Brits "were too wise to be counseled by American fishermen." Franklin thought a map of the current might help.

They were among the first to document the Gulf Stream's tremendous reach, charting it from Florida toward Europe. It was a massive river in the sea.

We know now that the Gulf Stream moves at a rate of thirty billion gallons per second, more than all the world's freshwater rivers combined. On its way, it hauls vast amounts of heat; a hurricane that twists into it gets a blast of fuel. It's a highway for migrating fish and a destination for deep-sea fishermen. It courses through an area that oil companies want to probe; an oil spill in the Gulf Stream would spread far and wide.

Though just sixty miles offshore from Charleston, the Gulf Stream has so much momentum it pulls water away from the coast. This whisking action actually lowers the sea level off Charleston like a downward pointing see-saw. If you could walk on water, a trek from the Gulf Stream to Folly Beach, one of Charleston's barrier islands, would go downhill three to five feet. Put another way, without the Gulf Stream's power, our tides would be at least three feet higher.

Skeptical? In 2009, sea levels from New York to New England rose five inches above normal. Scientists were stunned and later attributed the rise to strong winds that slowed the Gulf Stream and Atlantic's system of currents by 30 percent. The currents regained their strength about a year later, but scientists wondered: was this a blip? Has global warming somehow gummed up the currents? If so, what's next?

I dove into these questions and soon was swimming in stories about messages in bottles, abandoned telephone cables and undersea waterfalls. But below the surface was a larger one that the national media had all-but missed, one with tremendous stakes. A slowing Gulf Stream and associated Atlantic currents could rearrange marine life throughout the hemisphere. Sea levels on the East Coast would rise more quickly. Weather patterns from India to South Carolina to Scotland would change. Beyond these stakes, I learned that the race to understand the Gulf Stream also is about something else: the human drive to explore, that irresistible itch to look beyond what we know.

It could begin with Franklin's epiphany about the slow mail or on that increasingly bloody fishing boat I was stuck on that morning.

But why not start with Ringo?

And a yellow submarine.

And that song.

NASA officials survey the Ben Franklin research sub. Photo by NASA.

*We all live in a yellow submarine, yellow submarine, yellow submarine…*

It was summer, 1969, a time of inner and outer exploration, Woodstock and an Apollo moonshot. The Beatles' *Yellow Submarine* movie was out, with Ringo Starr doing lead vocals on its title song. Captain Don Kazimir picked up a cassette at a West Palm Beach mall. Catchy song. Perfect, really, given that he would soon pilot a real (partly) yellow submarine.

Captain Kazimir was in his mid-thirties, compact and wiry, with dark-rimmed glasses, more serious than swashbuckling, but his fascination with the sea ran deep. He was born in Ossining, New York, where his father worked at Sing Sing, the old prison on the wide Hudson River. Kazimir had always loved trips to beaches on Long Island. But, in fifth grade, his mother bought him Jules

Verne's *Twenty Thousand Leagues Under the Sea*, and his compass was set. Captain Nemo's tortured adventures? A battle with a giant squid?

Hooked, he later joined the Navy, became an officer and went to sub school in Connecticut. He played cat-and-mouse with Soviet submarines off Finland. He hoped someday to command his own ship, came close a few times, but Navy life is hard on families. So, he retired before reaching that goal and took a job with Grumman Aerospace Corp. Then, one day while paging through *Popular Science* magazine, his eyes landed on a story about the Gulf Stream Drift Mission.

The mission was Jacques Piccard's idea. Piccard was a famous Swiss explorer who, along with a Navy diver, had ridden a submersible 35,800 feet down into the Pacific Ocean's Mariana Trench, deeper than anyone else. Now Piccard had his sights on the Gulf Stream—a different challenge. No sea trench to reach; no immovable object to conquer. The Gulf Stream was a destination that moved. To truly explore it, you had to become part of it.

Piccard had somehow persuaded Grumman to build a submarine without any serious means of locomotion. It had four, twenty-five-horsepower motors, enough for minor maneuvers but too puny to move the sub's 140 tons out of trouble. It was fifty feet long and had twenty-nine portholes. It was painted yellow on top to make it more visible to surface ships, white on its underbelly. With its viewports and yellow-and-white paint scheme, it looked like a tube of Swiss cheese.

NASA signed on because the six-person crew would be sealed inside for a month, an experiment that would mimic living in space for an extended time. The Navy joined because observations might yield helpful data for submarine warfare. Piccard had noted: "if we are able to drift silently all along the American coast, several other submarines, maybe not American, will be able to do it, too."

Piccard needed a captain. Kazimir applied and finally won his first command.

The group christened their submarine the Ben Franklin because of his pioneering chart. And on July 14, 1969, a Navy ship towed the sub nineteen miles off Palm Beach, FL.

"Open the vents," Kazimir ordered. Water filled the ballast tanks.

Soon, the yellow submarine was under the waves, drifting.

---

The mission launched in what's generally considered the beginning of the Gulf Stream, the straits between the Bahamas and South Florida.

Here, currents from the Gulf of Mexico and the Caribbean tropics converge, forming a single hot flow. Pushed by trade winds, the Gulf Stream rolls north along the Florida coast.

Sixty miles wide, it gathers force as it passes Georgia and the Carolinas, its flow six thousand times greater than the Mississippi River. Past South Carolina, the Gulf Stream nearly collides with the Outer Banks, missing Cape Hatteras by just twelve miles. Then it jets northeast, surging deeper into the Atlantic. Away from land now, it undulates like a dropped fire hose. It spins off eddies that swirl in great arcs. Like long curls, the current and its eddies flow across the Atlantic toward northern Europe. Along the way, these curls carry heat equivalent to a million nuclear power plants.

The Gulf Stream can be fifteen to twenty degrees warmer than water outside it. During winter, steam wafts off it like smoke. During spring and summer, its heat rises and boils into massive thunderheads. One year, scientists noticed a line of clouds over it that stretched six hundred miles, as if a jet had created an immense contrail. In the North Atlantic, winds blow over the Gulf Stream and warm Iceland like a radiator. Farther south, it fuels tropical storms. In 1989, Hurricane Hugo crossed the warmth in the Gulf Stream. The stream gave the storm "a sudden shot of high octane," a meteorologist said at the time. Charleston wouldn't be the same for years.

When winds turn against the Gulf Stream, steep waves form, giant castles of water and foam. In 2005, a wave seventy feet tall struck the Norwegian Dawn cruise liner as it steamed from the Bahamas to New York. Waves broke over the bow, flooded sixty cabins and tossed debris as high as the ship's tenth deck. It limped into Charleston for repairs.

When caught in the Gulf Stream's flow, stuff and people go far. Between 2000 and 2007, researchers tossed 1,200 bottles off the Canadian coast, each with a message to report its location if found. The bottles landed on beaches from Finland to France.

In 1986, Nelson McIntosh lost power on his boat between the Bahamas and Florida. He drifted in and out of the Gulf Stream for about fifty days and more than four hundred miles, surviving on rainwater collected in his seat cushion. When fishermen rescued him off Charleston, he'd lost eighty pounds. His first request was a beer.

Though it has a name, the Gulf Stream doesn't really have a true beginning or end. That's because it's part of a larger and mostly hidden conveyor belt of currents, a system that keeps the planet's climate in check. And on this belt in 1969, the six crew members of the *Ben Franklin* floated north, picking up speed.

They spent hours gazing through the portholes. Off Cape Canaveral, they saw things out of Captain Nemo's logs: a squid attached to one of the viewports and a large colony of medusa jellyfish, their long tentacles wrapping around each other in a graceful embrace.

At night, the sub's lights lured plankton and other tiny organisms. Flashing glints of blue, yellow, and orange, they swirled like turns of a kaleidoscope. Because it seemed to fit the mood, Kazimir played the opera *Madame Butterfly* on his cassette deck, especially when the salps floated by—translucent sacs that expanded and contracted like lungs, then joined in long chains and twirled as if in a dance.

The sub seemed to become one with the Darwinian world around it, both predator and prey. As Frank Busby, a Navy oceanographer, looked out a porthole, a large swordfish attacked, ramming the sub just below the window, damaging neither sub nor fish. Squids squirted dark ink and bolted. Cuttlefish launched clouds of sepia. Drifting with the ink clouds, the crew watched them expand slowly like Rorschach tests.

Diving now, down two thousand feet to the seabed, another world appeared: skeletal crabs poking along a sandy bed, even a small ray. Despite the crushing water pressure, there's still so much life down here, Kazimir thought.

Another thought, a fear, sometimes washed through his mind: is the sub going to make it with all those portholes? At such depths, one bad weld or a crack could send water shooting in like a laser. The sub would implode in an instant.

Off Georgia, the sub began to rise and descend on its own, up and down two hundred feet in thirty minutes. They couldn't feel this motion because they were part of it, just as a passenger in a hot-air balloon feels no breeze. But they saw what was happening on their instruments: they were in the Gulf Stream's internal waves, great deep-water rollers that rise and fall like slow-motion breaths.

Tensions rose. The internal waves "are giving us fits," Kazimir wrote in his log. Temperatures inside the sub dropped to fifty-three degrees. Kazimir doled out a few mini-bottles of whisky and Scotch to warm bodies and spirits, a Navy tradition. They paid close attention to sonar readings. Uncharted shipwrecks could be disastrous, and they were near the last-known coordinates of a missing World War II tanker.

In 1943, a German U-boat fired torpedoes at the *Esso Gettysburg*. Bursting into flames, it sank with 131,000 barrels of oil. Years later, with the shipwreck still not found, the National Oceanic and Atmospheric Administration estimated that if 10 percent of this oil leaked into the Gulf Stream, the slick could affect an area the size of California.

Suddenly, off Georgia, the sonar picked up something ahead. Kazimir raised the sub from a depth of 750 feet to 100. It wasn't a shipwreck. It was much larger. A coral reef? Nothing on the charts.

The seabed was more complex now, a series of undulating scarps and caves. Hiding in the caves were great schools of wreckfish and squid. Those squid were "more beautiful than anything we had seen previously," the crew reported to the surface ship tracking them.

Some features rose four hundred feet, and oceanographers would later name this rugged area the "Charleston Bump." It deflects the Gulf Stream like a rock in a river rapid, creating eddies and turbulence.

On day eleven, the *Ben Franklin* found itself in one of those eddies circling back toward the South Carolina coast.

Captain Kazimir cranked the sub's tiny motors to move back into the current. But the motors were no match for the current's force.

Piccard wrote: "The Gulf Stream has rejected us."

---

Science is like a great current, a flow of ideas and surprises. It was especially so in 2009, when researchers watched in awe as the Atlantic's conveyor belt suddenly slowed as if something was caught in its gears.

Some had feared this might happen. Carbon dioxide levels in the atmosphere had spiked. The Earth's oceans and atmosphere also had warmed rapidly, causing Greenland's glaciers to melt at rates unseen in hundreds of years. This massive infusion of freshwater poured into areas near a critical point in that conveyor belt—the place where the Gulf Stream takes a deep dive.

Between Greenland and Norway, the Gulf Stream's warm and salty flow cools, which makes it dense and heavy. This heavier water then drops like a slow-moving waterfall. It sinks to the ocean floor, where it joins currents moving south. Scientists call this system the "AMOC," short for Atlantic Meridional Overturning Circulation.

A crimp in the conveyor belt would vibrate across the globe. With slower currents moving south, hotter places would get hotter. With less heat going north, winters in Europe would get colder. A sluggish AMOC could change monsoon cycles in Asia and South America. It would give tropical storms in the Caribbean more time to form and linger. This would lead to more downpours and stronger hurricanes in the Gulf of Mexico and Southeast. The sea level would rise along the East Coast.

But until 2004, scientists had just a few ways to measure this conveyor system. One was an abandoned telephone cable strung between Florida and the Bahamas. Through a nifty bit of physics, oceanographers figured out how to calculate the Gulf Stream's velocity by analyzing the line's voltage. But it was just a snapshot of one part of the conveyor belt, so scientists in the early 2000s launched RAPID-AMOC, an ambitious international program to string monitoring stations across the Atlantic.

And in 2009, these stations recorded a sudden 30 percent drop in the AMOC's flow.

"It was totally unexpected," said Harry L. Bryden, an oceanographer in England who'd organized the monitoring effort there.

"It was like nothing we'd seen before," said William E. Johns, an oceanographer in Miami who'd led from the American side.

As the flow diminished, weird things happened: Europe had record-breaking snowfalls; sea levels north of New York City rose five inches; the Gulf Stream twisted farther north; temperatures in the Gulf of Maine shot up faster than 99 percent of the world's oceans; codfish stocks there plummeted; Norfolk, Virginia's low-lying streets flooded more frequently on sunny days; Charleston's tides were higher than predicted.

A huge change in the world's underwater machinery was underway, hidden below the waves.

---

Rejected from the Gulf Stream off South Carolina, Captain Kazimir brought the sub to the surface, moving through brighter shades of green. In the sunlight, they saw diatoms, tiny plants that soak up carbon dioxide and rival snowflakes for their beauty and sparkle. They saw barracudas and hammerhead sharks.

They broke through the waves but kept the hatch sealed—kept their micro-climate intact to preserve NASA's research.

They'd already learned that six guys in a tube meant constant noise, and not much sleep. And Piccard wasn't impressed with the food: canned, freeze-dried and packaged stuff. Kazimir didn't mind. He was used to worse on Navy subs. But everyone agreed: bringing that dartboard was a great move. NASA later made sure one went up with the space station Skylab, though with tips made of something new called Velcro.

The sub's climate grew dirtier as time passed. The waste system struggled. The air stank. Two crew members developed a rash.

"Probably due to perspiration and the fact that underwear was changed every three days (not often enough)," Kazimir wrote in his log.

To keep the sub from turning into a greenhouse, they deployed rolls of silica gel to sop up the humidity. They hung them everywhere like deli sausages. They had plenty of oxygen; it came from liquid oxygen tanks. But $CO_2$ levels rose quickly. They monitored themselves for headaches, fatigue, and the shakes—signs of $CO_2$ poisoning. Panels of lithium hydroxide captured excess $CO_2$. When the crew changed them, powder filled the cabin and made everyone cough.

Drifting on the surface, the crew listened to CBS News Radio 88. The Apollo mission to the moon had just happened. One small step on the moon, a small step backward out of the Gulf Stream.

A Navy ship pulled aside and attached lines. The tow back into the Gulf Stream took seven hours, from green water to blue.

Released from the lines, the sub went back under.

Kazimir inserted a cassette: Willie Nelson. They were on the road again.

Science also is like the turbulent current Captain Kazimir and his crewmates were in off Charleston, with eddies of confusion and debate amid a stronger forward flow.

In 2010, the Atlantic conveyor belt regained its strength—almost. Its velocity was still weaker than before.

Scientists paddled harder to find out what happened and what it meant. They studied tidal gauges, wind patterns and data from that abandoned telephone line, the one strung between Florida and the Bahamas.

William Sweet, a NOAA oceanographer in Maryland, discovered that the slowdown came amid other changes in the weather. The combination generated higher tides from Georgia and Virginia, tides that pushed water into normally dry streets. "People can literally put their feet in the water and say it's related to the Gulf Stream offshore," he told me.

Tal Ezer, a scientist at Old Dominion University, uncovered evidence that hurricanes off Florida could disrupt the Gulf Stream for several days, triggering sunny-day floods hundreds of miles away in Norfolk, where the school is located.

Could a force as powerful as the Gulf Stream be so fickle?

Tom Rossby, a Rhode Island oceanographer, told me he was skeptical. He'd set up a program to measure the Gulf Stream by outfitting a freighter with instruments. The freighter had made regular runs between New Jersey and Bermuda for more than twenty years.

"I can happily report that the Gulf Stream isn't slowing down," he said. His direct measurements were proof. "It's really quite stable. Papers that say otherwise are baloney."

Over the next few years, though, Rossby's happy report was lost in a flood of new evidence. Teams of researchers looked for clues in cores and fossils; they gathered data from new monitoring stations in the North Atlantic—the place where the Gulf Stream does its deep dive. They ran programs on computers capable of doing quadrillions of math problems a second. Stable? The Gulf Stream?

Not if you looked at the entire conveyor belt, scientists including Harry L. Bryden countered. Then, in April 2018, *Nature*, the science journal, published two alarming studies.

Each study was based on years of research. One team used sediment cores from the ocean floor along the East Coast. The cores told you things about ocean temperatures in the past, just as the width of tree rings tell you when growing seasons long ago were good or bad.

The study found the Atlantic's conveyor belt began to slow in about 1850, weakening as the Little Ice Age ended. Researchers said that rapidly rising greenhouse gases may have amplified this natural weakening trend.

A second team analyzed sea temperature records and ran simulations on a super computer. For climate scientists, these computer models were like new high-definition TVs. Once-fuzzy images of the Atlantic's conveyor system suddenly grew clearer.

They found that amid a rapid increase of carbon dioxide levels, the Atlantic's current system had weakened since the 1950s. Researchers said a slower flow could have shifted the Gulf Stream farther north, created more storms in Europe, and worsened droughts in parts of Africa.

The studies had different methods but similar conclusions: the Atlantic Ocean conveyor belt had slowed by 15 percent.

By some measures, it was at its weakest point in sixteen hundred years.

Past Cape Hatteras, the Ben Franklin entered deep water, drifting with schools of migrating tuna. Past Virginia, then Maryland, and the Ben Franklin picked up speed, moving at five mph. Past New Jersey and Massachusetts, and Captain Kazimir wrote in his log, "the crew is getting restless."

Inside, carbon dioxide levels reached more than 13,000 parts per million. Crew members reported feeling short of breath at times. But the mission was nearing its end. Frank Busby, the Navy oceanographer and a heavy smoker, was desperate for a cigarette. To Kazimir, it seemed as if he'd had his bags packed for days.

On Day Thirty, Kazimir prepared to surface off Nova Scotia. They took last looks through the viewports as the sub rose in the brightening light: a jellyfish floated by; a chain of salps did their dance. By then, Chet May, a NASA engineer, had inserted that Beatles cassette for the last time.

*And we lived beneath the waves in our yellow submarine.*

By all accounts, the *Ben Franklin* mission was a success. The Gulf Stream had never been analyzed so deeply. Crew members made nine hundred thousand measurements of temperature, salinity and other features. They'd been surprised by the size and power of those internal waves, the turbulence off Charleston, the attacks by those foolhardy swordfish. They'd withstood thirty days in an unheated metal tube, a record for a civilian sub.

But the mission was largely lost to history. While the aquanauts of the *Ben Franklin* were under the waves, the Apollo astronauts walked on the moon. The world was looking up, not down, and after some time, the *Ben Franklin* ended up in a Canadian junkyard.

More years would pass before a maritime museum in Vancouver, British Columbia, discovered this bit of history hiding in plain sight. The museum restored the *Ben Franklin* and put it on display, touting "The Story of Our Yellow Submarine."

Today, Don Kazimir, in his late eighties, talks about the Ben Franklin expedition with the kind of urgency you hear near the end of long missions. He's one of two surviving crew members, recently retired from a Catholic charity, living near the water in South Florida on Gulfstream Road.

One afternoon, Kazmir ushered me into a den and pulled out boxes he'd stowed in a closet and his garage: *Ben Franklin* mission patches, his uniform, logbooks, charts. Memories resurfaced about those glittering plankton and ink-squirting squid. He found himself thinking hard about the mission's legacy, his place in it, Greenland's melting ice, the impacts of a slowing Gulf Stream—the importance of exploring all parts of Earth, a blue-and-green submarine drifting in space. "That's the curiosity built in all of us, to learn."

Meanwhile, across a lagoon by Kazimir's house, the Atlantic lapped onto an eroding beach. And, beyond the horizon, the blue Gulf Stream flowed, its power and changes hidden a little less now because of explorers in a yellow submarine.

A massive iceberg near Illilusat, Greenland.
Photo by Lauren Petracca of *The Post and Courier*.

# THE GREENLAND CONNECTION

SO MANY THINGS IN GREENLAND ARE GIGANTIC. GREENLAND IS five times the size of California, and roughly 80 percent is covered with ice. Greenland's ice sheet is a mile deep on average, but near the center of the country it rises ten thousand feet into the sky. Greenland's ice sheet is so thick and heavy that it makes the Earth wobble a bit as it spins, like an unbalanced top. When the ice sheet meets the ocean, the ice sometimes cracks and falls with the force of atomic bombs. Even Greenland's language, Greenlandic, has huge words—one is 153 letters long.

Greenland's ice is melting in a big way, too. During the summer of 2021, so much melted in one week that you could flood the entire state of South Carolina with two feet of water. The ice sheet normally melts in the summer, but it's melting faster now than it has in twelve thousand years.

All this melting ice raised sea levels across the globe, just as dropping ice cubes into a whisky drink eventually makes a mess. Except some ice cubes in Greenland can be half the size of Manhattan.

There's more: the Greenland ice sheet is so massive that it generates its own gravity. It pulls the Atlantic Ocean toward it like someone tugging a blanket. South Carolina is at the other end of this blanket, which means that Greenland pulls water away from our coast, lowering our sea level. But as the ice melts, its gravity disperses, and its grip loosens. Seas at the far end of the ice's power slosh back.

I had trouble wrapping my mind around the notion that Greenland somehow had a gravitational force of its own, one that affects how much my neighborhood in Charleston flooded. Greenland is more than three thousand miles away from South Carolina. Could such a faraway place of polar bears and reindeer really hold the key to Charleston's fate? There was something numbing about Greenland and its outsize role in climate discussions, something that applied to talk about climate change as well. Perhaps it was its distance and scale. So, during a three-week trip to this wonderland of ice, I tried to narrow my focus. Which is how I ended up on a seventy-eight-year-old plane flying over the world's fastest-moving glacier. With an Elvis impersonator on board.

---

It was the middle of August 2022, and the afternoon temperature in southwestern Greenland was in the low sixties, speeding up the summer melt. In the sky above, Josh Willis crouched in the back of a World War II-era DC-3.

He wore a blue NASA jumpsuit and cradled a metal tube that was about three feet long. He peeled off a sticker that said "REMOVE BEFORE LAUNCHING." Setting the tube down, he opened a round metal hatch in the floor. Through the hole, you could see the Ilulissat Icefjord below.

Josh has a cherubic face and those long Elvis sideburns. When I first mentioned that he really does look like Elvis, he lowered his voice and answered with the King's trademark, "Thank you very much." He's a graduate of Second City's comedy school in Los Angeles and has done shows on Hollywood Boulevard. He confessed that his performances are a bit of oil and water—climate science and comedy. But he thinks that scientists could do a better job talking about their discoveries, and humor helps. For a science communication contest a few years ago, he and friends did a music video called the "Climate Rock." In it, an eleven-year-old asks, "What is climate?" Josh, in a 1970s Elvis jumpsuit, sings: "you take a bunch of weather and you average it together and you're doing the Climate Rock!"

Climate Elvis was born.

Josh has a more serious day job: climate scientist with NASA's Jet Propulsion Laboratory in Pasadena, California. He leads the agency's OMG project, which does not stand for "Oh My God," though he does find himself saying that when he looks below and sees Greenland's cathedrals of ice. It stands for Oceans Melting Greenland, a title he cooked up a decade ago as a catchy way to describe the project's central question: do warming oceans affect Greenland's ice sheet?

Which is how he ended up throwing things out of airplanes.

The Ilulissat Glacier is a key OMG target and one the most important glaciers you've never heard of. It pours into a large valley near the town of Ilulissat, which is pronounced illoo-lih-sat and means "icebergs" in Greenlandic. The glacier also goes by other names: Jakobshavn, after a Danish merchant, and still used by many scientists; and the Greenlandic name "Sermeq Kujalleq," or south glacier. But given all the giant icebergs, Ilulissat fits best.

About forty miles from the sea, the Ilulissat Glacier forms an eight-mile wall called a calving front. Here, ice moves toward the ocean at 150 feet per day—a pace that tripled during the 1990s and 2000s. As it moves, it creates a great white shelf over the water that breaks off, often violently.

On warm days, the ice cracks like cubes after they've been dropped in a warm drink, except these cracks sound like thunderclaps and shake your ribs. Chunks as large as skyscrapers crash into the water, launching ice shards and spray. Some fractures release so much energy that geologists call them glacial quakes. Earthquake instruments across the world detect the biggest calving events. In 2008, a crew for the documentary *Chasing Ice* watched part of the ice wall collapse in a roar of thunder and white. The chunk was larger than three thousand Egyptian pyramids.

All this falling ice flows down a fjord that's 2,500 feet deep. But near the fjord's mouth, the biggest icebergs hit an underwater speed bump—a sudden rise in the seabed that's still about eight hundred feet deep. This bump creates the world's most beautiful traffic jam.

Icebergs with giant arches crowd ones that look like snow cones, alligator heads and cowboy hats. Blue meltwater rivers speed down shimmering white slopes. Humpback whales swim between iceberg cliffs. Water streams off the cliffs, sounding like a steady rain. Without warning, icebergs sometimes do somersaults, even ones the size of aircraft carriers. My colleague, Lauren Petracca, and I sat on a hill watching the icebergs one afternoon when something moved. Lauren had set up a timelapse camera. When we looked at it later, we saw what looked like an entire mountain flip over. These cartwheeling icebergs can swamp fishing boats and smear the water with white ice bits for miles.

Over time, ice melts below the big icebergs, enough to clear that eight-hundred-foot-deep speed bump. Freed from the fjord, they float into the open ocean, propelled now by powerful currents.

But this traffic jam had long given OMG fits. The NASA crew needed space in the water to drop their probes, and sometimes the bergs were bumper to bumper. A few days before, they'd found an opening to drop a probe. But it didn't broadcast any data. Now they were back for another try. And Josh Willis badly wanted the measurements, in part because of what they'd discovered a few years before.

The probes do two simple things: measure water temperature and saltiness. But they do this in a complex way; After Josh pushes the tube out of the plane, a parachute opens, slowing the probe to about sixty miles per hour before it hits the water. A battery that uses saltwater to generate a charge then triggers a second probe.

Tethered by an unspooling wire, the second part sinks toward the seabed, beaming up data to the plane. Satellites and aerial surveys can't measure temperature and salinity deep in the sea, Josh explained. "So, you have to put a thermometer in it." And they need to do this from a plane because of Greenland's size—nearly 1,700 miles from its northern tip to its southern, or roughly the distance between Charleston and Denver.

The OMG flights began in 2016. Flying low, they dropped about 250 probes around the entire country that year. Almost immediately, they stumbled upon something unexpected.

The Ilulissat Glacier was growing.

This was surprising because Ilulissat is known in scientific circles as a global floodgate. Floodgates are massive faucets that drain the ice sheet, and Ilulissat's melt is so formidable it may have already contributed more to sea rise than any other single feature north of Antarctica. But the OMG data showed it was expanding for the first time in twenty years. "At first blush, it seemed like great news," Josh said. It wasn't.

More readings showed that a large blob of cold water had moved into the fjord, temporarily cooling the glacier like a big ice blanket. This allowed the ice above and below to grow. The glacier expanded again in 2017, 2018, and 2019.

But in 2020, the cold water left the fjord. Warm water replaced it, and the glacier shrank. Bad news for places like Charleston, New Orleans and other cities at or near sea level. For the OMG scientists, "it was a home run," Josh said. It proved their hypothesis: what's happening hidden below the waves affected when and how glaciers melted. And it showed that the warming waters of the Atlantic and its undulating currents drove that melt. Based on this OMG work, scientists were able to calculate that Greenland's already major contribution to sea rise might be twice as large as previously thought.

As Josh readied the probe, summer heat waves had smashed records in the western United States and Europe. Josh wondered: was the water in the Ilulissat Icefjord still getting warmer?

In the cockpit, a Canadian pilot named Jim Haffey saw a few openings in the ice floe. I thought about his very thorough pre-flight briefing. ("If the plane crashes and everyone is unconscious, here's how you turn off the engines.") He put the old plane in a hard left turn. "This could work," he said, motioning to a patch about the size of a basketball court.

"I like it," said Mike Wood, a post-doctoral researcher at NASA's Jet Propulsion Laboratory.

"Let's give it a shot," Josh said, loading the probe into the chute.

The old plane dove. The icebergs grew larger. The altimeter ticked below seven hundred feet.

Six hundred.

Five hundred.

"Now!" the pilot said.

"Twelve away," Josh said, pushing the probe down the chute.

The pilot banked left hard, looking for the probe's splash.

In the back, the scientists waited for a signal.

Mike Wood, left, and Josh WIllis of NASA prepare to launch a temperature probe during their Oceans Melting Greenland (OMG) work in 2022. Photo by Lauren Petracca of *The Post and Courier.*

That same weekend, three thousand miles south, Hurricane Grace churned toward Veracruz, Mexico—the strongest hurricane ever recorded in that region. Tropical Storm Fred pounded the Florida Panhandle, flooding areas with seven to nine inches of rain as it moved inland toward the Carolinas. Henri formed off the East Coast, beginning its destructive trek to Rhode Island.

You can't pin climate change as the cause of these storms, but physics and patterns signal they're getting worse. The average global temperature has risen 1.8 degrees Fahrenheit since 1975, and physicists know that every increase like that means the air can hold 7 percent more moisture. Put another way: in a warming world, the buckets above our heads get larger.

At the same time, ocean temperatures are rising, up 1.5 degrees since 1901. Warmer water naturally expands. And more volume means higher sea levels, more record-breaking floods. And all this warmer air and water fuels more intense storms. Hurricanes are intensifying more rapidly; they're forming earlier in the season; they're dumping more rain. In 2020, the overheated waters of the Atlantic spawned a record thirty storms, so many that forecasters began using Greek letters to name storms.

In the Southeastern United States, this extra heat translates into a 27 percent increase in torrential rainstorms—summertime downpours that would soon strike coastal South Carolina.

---

But before that happens, let's get out of the sweatbox, head back to the Arctic, where flying over the Greenland ice sheet is one way to get a sense of things. Another way is to go even smaller. Put boots with metal crampons onto the ice sheet itself as it melts, hear the clomp, clomp, clomp of your guide, Adam Lyberth, as he steps toward the edge of a blue-and-white chasm that could be your last memory if you slipped.

The ice itself was hard but soppy. Walking on it, I spotted more small cracks than I expected, like breaks in tempered glass. But it was the blue hues around us that captured my imagination. Set against the white, the blues were so vibrant they seemed at once pure and fake. The scientists who dreamed up toilet bowl cleaners nailed the same tint.

Adam Lyberth explained that bubbles in the ice make it white. When layers of snow pile up over centuries, those bubbles get squeezed out, and the ice takes on a sapphire shade. This blue ice is denser and safer to walk on. You also see this color in meltwater lakes and streams. The streams curl through the ice sheet. They pour into fissures that drop for hundreds of feet. Gurgling streams rush under ice bridges. They carve deep canyons over time, making formations like those in southern Utah—except everything here is wet and white and blue with streaks of black. You couldn't feel it, but all that water and ice under you was moving toward the ocean.

Adam grew up in one of the small settlements that cling to the coast in Western Greenland, not far from the town of Kangerlussuaq. Now, a day before his sixtieth birthday, he said his present was "to be here on the big ice." He said he's visited the big ice at least three thousand times, mainly as a guide for tourists and European royalty, taking them to places "where if you fall …" He paused and waved bye-bye with his hands. Adam is a thoughtful man, and as a shaman elder, he said he's experienced nature in ways that may seem foreign even to other native Greenlanders; he sometimes sees shades of light move past his field of vision that portend calamities. He can do an impressive musk ox call. Nature helps connect our minds with our hearts, he said. But over the decades, he's witnessed a disconnect in the nature around him, the kinds of changes that Josh Willis and Mike Wood from NASA and so many other scientists were trying to measure.

Along the ice sheet's edges, you could see land where giant ice walls rose a few years ago. The ice that's left melts faster, and it's dirtier, Adam said. Some of the dirt is wind-borne dust left by melting glaciers. And some is black carbon, the soot-like specks from distant wildfires, vehicle tailpipes, coal plants, even microparticles from tires as they wear down. These dark particles absorb more heat than white snow. The evidence was by his feet. Small-bore holes—places where dark specks landed and melted into the ice.

It's warmer, he said. Winter temperatures rose eight degrees during the past thirty years, and summer temperatures were three degrees higher. Warmer weather thawed the permafrost, making buildings from Greenland to Alaska sink and tilt. Melting ice exposes land that had been covered for thousands of years, releasing gases that reminded him of a pig farm. And Greenland usually gets relatively little snow and rain, Adam added. But warmer air holds more moisture.

"The rain falls harder now."

We hiked for an hour deeper into this wonderland, and then it began to rain. Adam moved more quickly, sensing danger. But he wanted to show us something before we turned around—the lip of an ice canyon. We made it to the edge. Below us was a Grand Canyon of ice that stretched toward the horizon. The sun suddenly broke through, and the ice sheet sparkled for a moment. The meltwater gave everything a watery sheen.

There, on the edge, I asked Adam what he thought about the rapidly melting ice, and about what was happening elsewhere: the wildfires in Europe and the United States, the record-breaking heat waves, the changes in the Arctic.

He told me these changes were spiritual warnings as much as climatic ones. "I think the heart of man has melted."

Adam Lyberth, an Inuit shaman and guide, looks out onto the Greenland ice sheet. Photo by Lauren Petracca of *The Post and Courier*.

Later, I looked at the weather in Charleston. The air there was heavy and wet. Weather balloons launched by the Charleston International Airport recorded enormous amounts of water overhead; the buckets were full, and it wouldn't take much to tip them over.

I imagined what would happen next. My neighbors would rush to move their cars. A woman in her 70s would likely drag a metal barricade across the street to prevent motorists from plowing through. A wake by a truck once generated a wave so high it wrecked her car. Ambulances on the way to downtown hospitals would detour around low spots to avoid being swamped.

Rain is fairly rare in Greenland. What were the chances it would pour in Charleston and Greenland on the same day?

---

High above Greenland, the ice sheet looks like a vast flat Sahara. And standing on the ice with an Inuit shaman reveals its more complex contours. But I work for a newspaper in Charleston. Why should readers there care?

Here, it's best to step back again from Greenland for a moment and think about bathtubs.

For many years, scientists assumed that just like turning on a faucet raises the water level in a tub, melting ice in Greenland and Antarctica raised sea levels uniformly across the globe. Scientists predictably called this "the bathtub model."

But the bathtub model never explained why some places, including the Southeast and Gulf coasts, had higher rates of sea rise than others. Subsidence and the lingering effects of the Ice Age partly explained these differences: some areas naturally sink as Earth's tectonic plates move about. But when scientists factored out subsidence and the Ice Age, some coastlines still had higher rates.

Then, in the late 1990s, a team led by a scientist named Jerry Mitrovica began thinking hard about Greenland's huge ice sheet. As physicists, they knew Greenland's ice locked up gravity. And gravity's ability to move the seas was as obvious as a high tide. Greenland's heavy ice sheet also must pull the ocean toward it, like a miniature moon, they thought. Like an incoming lunar tide, this should raise sea levels near Greenland.

But on distant shorelines, this gravitational force would do something else: pull water away from coastlines, lowering their sea levels like an outgoing tide. Mitrovica and his colleagues wondered: what if the Greenland ice sheet melted?

So, using computer models, they filled the bathtub, simulating what would happen if enough ice in Greenland melted to raise global sea levels one meter. Then, taking changes in gravity into account, they mapped the results. "That was the eureka moment," Mitrovica, now at Harvard, told me.

There, in red and blue hues, were Greenland's long gravity fingerprints. In Mitrovica's modeling, less ice in Greenland meant lower sea levels around Greenland as the diminished gravity pulled less water toward the island. But at about the 1,200 mile mark, seas rose as the oceans surged back.

At the time, Mitrovica tried to explain these counterintuitive findings to a group in the Netherlands. "They practically ushered me out of the room, thinking I was a lunatic."

But their work would provide the foundation for our understanding of a key but underappreciated force that affects hundreds of millions of coastal residents. And, more recently, NASA scientists took the gravity fingerprint concept a step further.

They looked at which areas of melting ice in the world affected which seaside cities the most, including Charleston. In other words, how much did ice loss in Greenland affect Charleston versus glaciers in Alaska or Antarctica?

The results showed that Greenland's melting ice had by far the biggest gravitational impact on Charleston's sea level—more than Antarctica and much more than all the world's melting mountain glaciers.

It showed that gravitational losses of polar ice during the next ten years would add half an inch of sea rise here, 40 percent of it coming from Greenland.

And Western Greenland was Charleston's gravity-loss hotspot, namely the area around the Ilulissat Icefjord.

The same place Climate Elvis was trying to probe.

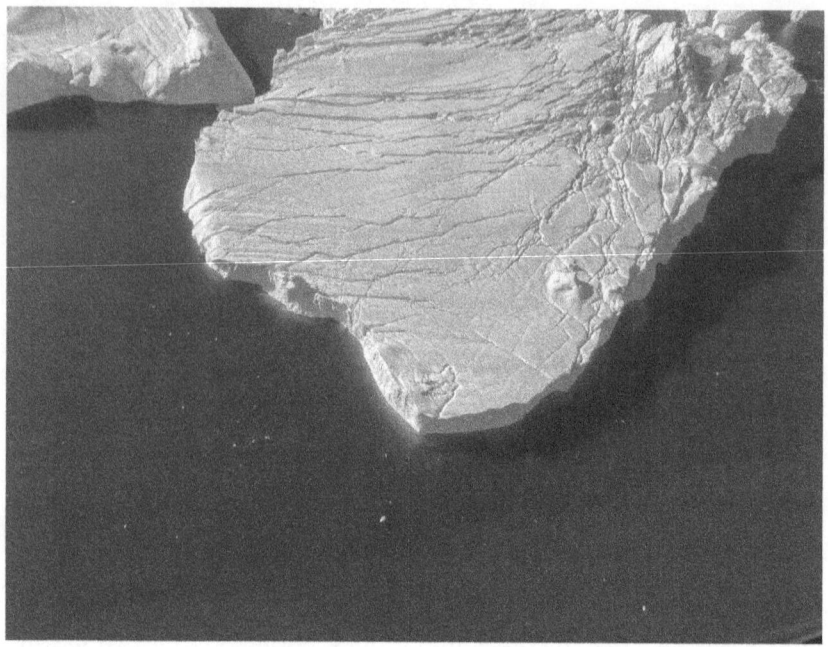

Icebergs fill Greenland's spectacular Ilulissat Icefjord.
Photo by Tony Bartelme of *The Post and Courier*.

The old DC-3's two propellers groaned as the plane circled above the Ilulissat Glacier. The mile-thick ice hides a feature that worries scientists like Josh Willis. Underneath, water extends another 189 miles into the belly of the ice dome. In other words, if the ice sheet was a bottle, the Ilulissat Glacier would form the neck. And that bottle will drain faster with warmer water below and warmer above.

"The cork is out of the bottle," Josh said.

A very large bottle.

Greenland's ice dome locks up a huge amount of freshwater, roughly as much as all the rivers and lakes in the Northern Hemisphere combined, including the Great Lakes—half of the freshwater this side of the equator. But Greenland's glaciers are melting six to seven times faster today than twenty-five years ago, with upward of five trillion tons of ice going into the oceans since then.

That melted ice alone raised seas by nearly half an inch. That doesn't sound like much, but scientists estimate that every half inch of sea rise means 6 million additional people around the world experience regular floods—a situation that residents in low-lying areas of Charleston, Miami, Norfolk, Virginia, and other coastal cities know only too well.

Some of this meltwater also poured into the AMOC, the conveyer belt of ocean currents in the Atlantic that I'd learned in my earlier stories about the Gulf Stream. Greenland's massive ice melt has tossed a giant wrench into this important system. As I'd learned before, a slower AMOC meant higher sea levels along the East Coast.

"The gravity issue may represent a small increase in Charleston's sea level, let's say 20 percent, and ocean currents might be another 20 percent, and thermal expansion (of the oceans) another 20 percent, but once you add up these and other 20 percents out there, you have a problem," Josh said.

Manning the monitors, Mike Wood heard the probe's signal, which sounded like an old fax modem. Was there enough saltwater to trigger the battery? The weather had been unusually warm, and it had been raining, as well. A few minutes passed. The pilot circled. No data. Another dud.

"Jakobshavn is being complicated this year," Mike Wood said, referring to the Ilulissat Glacier's Danish name.

"Jakobshavn is being a jerk," Josh said.

Greenland is gigantic, and so are the forces melting its ice, staggering at times. In between probes, I asked Josh how he navigated the winds of doom that howl through climate journals?

The numbing data? The frustration over fixed beliefs that prevent action? The troublesome experiments like today's?

"I gravitate toward comedy, because that's what I like," Josh said. "But I also wanted to do comedy about climate science because I think it opens a door when you share a laugh together." Laughing together makes it harder for people to treat each other like enemies, he said. It melts our hearts for a moment, but in a good way.

A few minutes later, they launch a second probe. And a third. Both duds. Just too much freshwater.

But three weeks later, they went back for one more try. This time, the probe worked. Josh and his colleagues will now crunch the numbers—possibly for the last time. NASA originally designed and funded OMG as a five-year experiment. The agency extended it a year, in part because of its findings last year that the Ilulissat floodgate was open again.

The DC-3 headed back to the landing strip, crossing over the ice sheet. It was cloudy, which helped bring out the blues in the meltwater lakes and rivers flowing toward the coast. Before we landed, Josh told me: "after OMG ends, Climate Elvis will be available for weddings and parties."

---

Back on the ground, I checked Charleston's weather again. A different kind of floodgate opened there.

The rain began to fall, and then it came down harder. At the Charleston International Airport, about 2.8 inches fell in a single hour. More than 3.8 inches fell at the Goose Creek Reservoir. Forecasters issued flood advisories. Water filled the streets and began to pool.

Another drenching rain. Another record for the day—4.1 inches, besting one set in 1971.

It was nothing, though, compared to what happened in Greenland that day. There, on Greenland's great ice sheet, three times the size of Texas, at the very top of its ten thousand foot ice dome, a science station's thermometer went above freezing.

Until recent years, this had happened just six times in the past two thousand years, ice core records showed.

In the past decade, it already went above freezing twice, and this marked the third time.

Even so, the station had never recorded any rain, only snow.

But that day, for the first time, it didn't just rain. It came down hard.

Like a Lowcountry storm.

An eastern black rail in a Lowcountry marsh.
Photo by Christy Hand of the SC Department of Natural Resources.

# GHOST BIRD

THE SUN SETS OVER A SECRET SPOT IN A SOUTH CAROLINA marsh, casting amber light on the grass. At this twilight angle, the sunbeams add extra green to the blades, which are as high as my shoulders. I watch the grass sway in a breeze gentle enough for dragonflies to land. They bounce on the tips, their wings glint in the softening sun, and, for a moment, my path looks as if it's filled with tiny mirrors.

A little black bird may be here, underneath these bouncing dragonflies, somewhere in these sparkling green waves. A rare bird called the eastern black rail. A bird so difficult to see that John James Audubon never saw one in the wild. A bird so stealthy that even the most ardent birders haven't seen one, though they may have heard their calls. So rare that Christy Hand, a biologist, asks me—no, pleads—that I not reveal this place to anyone. She knows mysteries are irresistible.

A black rail? Here? She knows that if word gets out the birding websites will light up. The birders will head to this quiet and beautiful place. They could harm the birds she loves and studies so closely. They could trample the nests, even though it's breeding season. They would come because it's rare, because they could check it off their lists, because so few black rails are left they may never get another chance. And what's to stop them?

There are fewer eastern black rails than some endangered species, but the federal Endangered Species Act doesn't cover them. Rising sea levels are snatching away their breeding grounds. Unless something happens, black rails could disappear altogether in a generation or two.

So, as we walk deeper into the marsh, looking for this ghost bird, I know Christy Hand is torn.

Yes, she wants everyone to know the black rail's story, a story that's much bigger than one about a little bird with bright red eyes.

But no, maybe it's better not to talk. Maybe it's best to keep these mysteries secret.

---

Right, it's easier to keep studying these phantoms in a stealthy way, like the black rail itself. Besides, Christy is more comfortable in the field than in the spotlight. She's a biologist with the South Carolina Department of Natural Resources, in her late thirties. She has a monk's stamina when standing motionless in a marsh as mosquitoes swirl about. She has a diviner's touch when placing motion-sensing cameras in the grassy tunnels where black rails nest. Few people have seen an eastern black rail, and fewer still have photographed them. But during the past five years, Christy and her colleagues captured more than thirty thousand photos and videos of black rails.

Okay, it's not the most photogenic bird. It doesn't have the swiveling head of an owl or the brilliant red plumage of a cardinal. It's a reluctant flyer, this bird. It spends much of its time running through spaces between the stalks. Because of this, scientists have called it a "feathered mouse." It's slightly bigger than a mouse, though, about four to six inches long. At 1.2 ounces, it's lighter than a golf ball. But take a closer look and its features grow on you.

Its pinprick red eyes are set above an obsidian-colored beak shaped like a shark tooth. It has black and gray feathers with hints of brown near the neck, like the color of mud and decaying marsh grass. It has speckled white spots on its wings, like the dappled light on Spartina roots. It's designed to blend in, to be there without you knowing it.

Which is why most birders never set eyes on them. Instead, they hunt for their calls, which in birding circles counts as a sighting. If you're lucky, you might be a few feet away when you hear a soft *grr*, a growl that may be a defensive call. Or its sharp *churt*. Or the more dramatic *kickee-doo!*

Early in her research, Christy spent nearly a year hiking into marshes, surveying, listening, looking, waiting. The best times to hear black rails are in late evening and dawn. So that's when she did her surveys, day after day, hour after hour. She didn't hear a peep. There in the marsh, rational and irrational thoughts hung like mist in her mind. Why didn't she hear them? Did she have a hearing problem? Was she in the wrong place? Did they still exist in South Carolina? Were they real?

She'd long been fascinated by birds and other animals. She grew up in Indiana, in a neighborhood by a pond. She loved watching the birds come and go. She liked frogs, too, and guided them to hidden places because neighborhood boys tried to stuff firecrackers in their mouths. Her grandparents gave her a *Reader's Digest* book about birds, and soon she was trying to identify the birds around the pond. One day, like a tourist in a loud T-shirt, a bufflehead duck appeared, a rare visitor for the pond, its head the phosphorescent colors of a peacock feather. She formed an animal club in school, earned a biology degree from Earlham College in Indiana, a master's from Clemson, did research in Alaska, and eventually landed a job at the South Carolina Department of Natural Resources.

Then, in 2013, one of her mentors at the Department of Natural Resources walked into her office and asked what she knew about black rails.

"What's a black rail?" she answered.

Not long after that, a supervisor said the US Fish and Wildlife Service wanted to know more about the presence of black rails in South Carolina. Would she help? At the time, a conservation group had organized a black rail conference in Savannah. It was a perfect chance to get up to speed.

The meeting changed her life.

She didn't know much about black rails, but the scientific community didn't either. Where and when did black rails breed? What were their migration patterns? No one knew for sure—the bird was just that secretive. But researchers did know black rails were in trouble.

They live on spongy land, wetlands with enough water to keep forests from taking over but not so much that their nests flood. This means they live on the thinnest of edges, edges growing thinner by the year.

Wetlands have long been under assault by human development, drained and dredged, farmed and ditched. More than half of our wetlands have disappeared since Europeans settled here. Now, rising sea levels are squeezing what's left and squeezing out black rails. Eastern black rails once were found from New England to Florida. Now, they're all but gone. Virginia and Maryland were strongholds. "We haven't heard one in Virginia since 2017, and Maryland's in the same situation," Bryan Watts, director of The Center of Conservation Biology, told me.

In the mid-1970s, you might hear seventy in one spot in North Carolina's Outer Banks. A recent survey found four.

Conservationists tried to sound the alarms. In 2010, environmental groups asked the federal government to protect black rails under the Endangered Species Act. After a court fight, the Fish and Wildlife Service agreed to study the black rail's vulnerability.

Christy heard about this research during that workshop in Savannah in 2013. She remembered hearing how researchers estimated that between 1,000 and 2,200 birds were left. A few years later, they cut that number in half. Was the black rail another climate change casualty? Something inside Christy Hand's mind began to shift. This little bird was an "indicator" species. Its struggles reflected larger threats to our wetlands. She suddenly wanted to learn even more about that bird, especially since South Carolina had an unusual feature: impoundments built by enslaved people hundreds of years ago, wetlands that might be the bird's last hope here.

Maybe, she thought, if she understood some of the bird's secrets, where they nested, how they bred and moved about, maybe this information could help us save them.

Yes, the more information out there, the better. Hadn't secrecy harmed black rails? Made them targets? From the beginning?

In 1836, a naturalist in Philadelphia discovered a black rail on his farm, the first sighting in America. He sent a specimen to Audubon, the ornithologist and painter. Audubon painted an adult and chick running toward a puddle. He bragged that the dead specimen "will swell my catalog to the number of 475."

Audubon never saw one in the wild, and the black rail's reputation for elusiveness only grew. Sightings generated excitement among bird collectors across the country, including Arthur T. Wayne, one of South Carolina's early and most revered birders.

Wayne lived in Charleston in the late 1800s and early 1900s. Part scientist, part entrepreneur, he made a living as an ornithologist who supplied bird skins to museums and private collectors. He added forty-five new species to the state's list and wrote 125 scientific papers, naturalist David Chamberlain wrote in *The Chat*, a publication by the Carolina Bird Club. On the business side, rare birds brought in more money, so Wayne hunted for the colorful Carolina Parakeet, which would soon be extinct. Ivory billed woodpeckers were worth $20 apiece, about $520 in today's dollars. They were worth much more than Bachman's warblers, also teetering on the edge and only worth $2.50, or about $70 today. In 1903, a boy brought Wayne an unusual egg from a nearby oat field.

"I hastened with gun and collecting basket," he later wrote.

He found the rail on a nest with eight eggs. Accentuating his find with italics, he wrote: "It can be readily imagined with what pleasure I saw the parent incubating the eggs, as I was the *first* person who had *ever* seen this secretive bird *actually* on her nest!" The bird raced into the cover of the oats—so fast that it reminded him of a scurrying field mouse. He heard a *kickee-doo* and flushed it. Then he shot it, stuffed it, and gave it to a collector.

Since then, the black rail's mystique has only grown. In 2010, a birder spotted a black rail in a wildlife refuge in Massachusetts. He posted his discovery on eBird, the world's most popular bird sighting platform. Within a day, thirty-four birders flocked to the spot. Two days later, fifty people showed up; four days later, another thirty, all straining to hear that *kickee-doo!* Black rails haven't been seen in Massachusetts since then.

Lewis Burke, president of the Carolina Bird Club, said his closest friend and mentor Dan Hudson spotted seven hundred bird species in the United States but never saw a black rail. Burke was luckier and heard one in a South Carolina wildlife refuge. He saw the marsh grass move but never saw the rail. "I nearly had a heart attack because I knew what it was." Peter Kleinhenz, a Florida birder, calls a black rail sighting akin to a climber summiting Mount Everest.

In 2014, as Christy Hand did more research in the marsh, she wondered if she would ever hear that call. She thought about the black rails' small place in the world, the long odds of their survival. Then, one morning before dawn in late June 2014, she and two colleagues went into the field again. It was humid, with a light breeze. The mosquitoes buzzed as the sun colored the sky orange. One of her colleagues played a few recorded calls, the *grr* and the *kickee-doo*.

And it happened. *Kickee-doo!*

Then another one in a different spot. *Kickee-doo!*

And she sat there, still, to hold the moment, listening to the birds but thoughts pouring through her mind: yes, her ears were fine. Yes, the birds were here. Later that morning, she noticed that her face ached from smiling so much.

Yes, she should talk about the black rail. More people need to know about the wonder of birds. If they learn about them, they'll love them, and we protect what we love, right?

Many of us already love birding. It's among the country's most popular hobbies. A study in 2016 estimated that the United States had 45 million birdwatchers, one in five. Half a million South

Carolinians count birding as a hobby. American birders spend a staggering 40 billion dollars a year on trips, food, and equipment. In terms of consumer spending, birding is bigger than hunting and comparable to recreational fishing.

The birding boom only grew during the coronavirus pandemic. With quiet streets, people suddenly could hear birdsongs they hadn't noticed before. Submissions to eBird shot up 46 percent in early April, when many states had shutdowns. On May 9, 2020, birdwatchers reported a record 2.1 million sightings on eBird for "Global Big Day," an annual bird-spotting event.

Ebird has been a big part of the birding boom. The Cornell Lab of Ornithology and Audubon Society created eBird in 2002. It's a free web-based platform that allows birders to crowdsource their observations. It takes in millions of sightings a year from across the world. It dangles incentives to stay involved; birders publish "life lists" of birds they've seen; top birders are ranked for all to see. It's fantasy football with beaks and wings.

This documentation happens in real time. If someone spots a painted bunting in a nearby park, you know exactly where to go.

Ebird can be addictive, avid birders say. Ted Floyd, editor of the American Birding Association's magazine, wrote in 2011 that eBird changed his life. "I keep at it, quite simply and quite profoundly, because eBird keeps me going, because eBird sustains me."

A recent look at eBird showed that people observed black rails 5,713 times, a comparatively small number. The endangered red-cockaded woodpecker had more than 19,600 observations, while common species, such as the Great Blue Heron, had 5.4 million and counting.

But even before eBird, a subset of birders took extraordinary lengths to check birds off their lists. These competitive listers fly to remote parts of Alaska to see rare species. They charter helicopters to spot birds in rugged mountains. All of this is done on the honor system, though some people cheat. They're called "stringers" because they string others along with sightings no one else confirms, said Nate Dias, an avid birder in Charleston who sits near the top of South Carolina's eBird lists.

The Olympic marathon of listing is a "Big Year"—counting the number of species in North America in a single year. It's a loose competition. A man named Olaf Danielson did a Big Year naked, birding mainly in nudist colonies and later writing a book called *Boobies, Peckers and Tits*. The 1998 Big Year was notable. Three birders traveled more than 275,000 miles to build their lists, a race chronicled by Mark Obmascik in his book *The Big Year* and later turned into a movie starring Jack Black, Steve Martin, and Owen Wilson. Sandy Komito, the winner that year, identified 745 species. In one scene in the book, Obmascik described Komito's hunt for yellow rails in the Anahuac National Wildlife Refuge in Texas. Yellow rails are just as secretive as black rails but more numerous.

"The only way to see yellow rails was to force them to take wing," Obmascik wrote.

Federal officials once obliged birders by driving through marshes in vehicles with tires you might see on monster trucks. This flushed the rails. Wildlife officers eventually stopped these destructive truck trips. In their place, Obmascik wrote, the birders still had their "secret weapon: terror." Komito and van loads of birders converged on the refuge. They had long ropes and plastic milk cartons filled with gravel. Attaching the cartons to the ropes, the birders formed a line and dragged the ropes and cartons through the marsh. The commotion righttened a yellow rail into flying. The birders checked it off their lists.

---

No, maybe Christy will keep quiet about black rails. Her work could stoke more curiosity about them and where they might be, trigger a rail rush. Then again, other people will gladly talk about them.

Like Dennis Forsythe, emeritus biology professor at The Citadel, the military college in Charleston. He's perched atop eBird's South Carolina list with 405 species. Years ago, Forsythe saw a black rail in eastern North Carolina. He was a partner in a bird tour company that took a group to a location on the Outer Banks. They heard a call, formed a circle, and flushed one. He couldn't believe his luck. It was there and gone so fast that someone in the group wanted another look—not realizing the rarity of a sighting.

Nate Dias, the Charleston ornithologist and conservationist, won't say exactly where, but he's seen black rails in the Santee Delta and ACE Basin—places where enslaved people once built hundreds of miles of dikes to grow rice. Many of these old rice-growing impoundments still exist, often as duck-hunting preserves. The dikes allow property managers to control water levels, in some cases creating that thin film of standing water that black rails love. "I fear that in ten years these rice field impoundments may be the only places you find black rails in South Carolina."

And there's Drew Lanham. Like Christy Hand, Lanham loved birds as a child, their colors, their freedom to get above everything. Sometimes in Edgefield, SC, where he grew up, he lay face-up on a dirt lane to watch buzzards trace circles in the sky. "I often wished we could trade places, that I could sail as effortlessly on the wind as they did," he wrote in his memoir *The Home Place: Memoirs of a Colored Man's Love Affair with Nature.*

Motionless, he waited until the buzzards were so close he could almost count the feathers on their wings. He went to Clemson on an engineering scholarship, then switched majors to follow his love for nature. He eventually became a professor of wildlife ecology at Clemson and a rarity himself, a Black birder. "Any bird that's black is my bird," he says on a YouTube video.

He was in his early twenties when he had his first encounter with black rails. It was on a trip to Fairlawn Plantation, a tract in the Francis Marion National Forest near Awendaw. Fairlawn has a long history of rare bird sightings, including the Bachman's warbler, which hasn't been seen anywhere since 1988 and may be extinct. He was with one of his mentors one night, and they had just left the property's old wooden cabins. They went into the marsh "and we heard this *kickee-doo, kickee-doo,* and they got louder and louder until it sounded almost like a chorus of frogs."

Back then, in the 1980s, the bird was known to be elusive but not on the edge of extinction, he said. But sometime in the 1990s black rails reached a tipping point. Populations in Maryland and Virginia collapsed.

These areas also saw some of the nation's highest rates of sea level rise. Part of this was driven by climate change, but land here also is sinking, a process known as subsidence. Some of this subsidence happens because of the natural movement of the planet's plates. Groundwater extraction is another contributor, and it's a serious problem in coastal areas of Maryland, Virginia and the Carolinas. Removing groundwater causes soils to compress—for the sponge to flatten. Seas rush in. But the old rice impoundments in South Carolina? They're safe havens, Lanham said. And haunting ones.

Now, when he visits the remnants of these rice fields, he finds his mind moving back and forth, from past to present, to the ghosts of enslaved people to the black rails—ghosted out of so many other places by rising tides but running and flying free behind dikes once used to make planters rich.

---

Impoundments. Rice fields. As a journalist, am I giving too many clues? Even federal officials won't talk about where black rails live. In late 2018, the Fish and Wildlife Service published a proposal to list the eastern black rail as a "threatened" species, one step below "endangered." It did so despite its own finding that without urgent action, the black rail will be extinct in thirty-five years. The agency had one year to greenlight the plan or put it aside.

Typically, proposals identify specific places where a teetering species might live. Not so for black rails. The agency noted the eastern black rail's "grail-like status in the birding community." It warned how birders use eBird to find black rails and converge at the mention of sightings. Carrying smartphones loaded with black rail recordings, they could broadcast calls in the marsh, disturbing the birds, causing them to flee their nests. They could trample nests, making rails more vulnerable to predators. You can find nuclear missile silos on Google Maps, but finding black rail habitat may be more difficult.

The Fish and Wildlife Service's proposal was the culmination of years of research, including Christy Hand's many discoveries in South Carolina. When she began in 2013, her first priorities were to learn where they nested and how many were left. Documented sightings were scarce, but she had a head start: a research project by a Clemson graduate student. That project identified seventeen places in South Carolina where observers heard black rails; they never saw any. Building on that work, Christy and colleagues did another survey over the next two years.

"I'd hoped that we'd find black rails sprinkled all over the place. But we didn't," she said. Most were still in or near those old rice fields.

Her next goal was to learn more about their behavior and breeding. She thought about attaching transmitters. But the birds were too small, and the dense marsh grass made it tough to pick up signals. She managed to trap and band a few birds, but that was risky, and she worried about harming them. Once, she caught a black rail with a net. Holding it, she noticed its eyes close, as if giving up—a sign of stress. She immediately let it go.

She decided on a new and less invasive plan: station cameras in the marsh. In an ocean of grass, she developed a knack for finding spots where the rails moved about. These cameras captured just a foot or two of space in this green and brown sea. But over time, she compiled the world's most comprehensive image and audio database on eastern black rails.

As she added to her database, she watched them grow. At first, newly hatched chicks had short and wobbly legs; chicks stumbled and used their wings for balance; after two weeks, their feathers emerged; by twenty-eight days, their bills were black; by day forty-two, their feathers were good enough for flying. With these images, she learned how they courted: males often offered females food, preened or simply chased after their mates. After coupling, a male tumbled from the female and circled with a lowered head and raised wings. The female bowed and ruffled her feathers. These videos also captured the sweet chirps of baby rails, rolling, high-frequency peeps amid their parents' *kickee-doos!* Smiling, Christy watched chicks follow their parents through the grass stalks. And these birds were sneaky.

Sometimes one approached her with a growl. Then, a few moments later, a bird growled from behind her. For all she knew, it might have run between her legs. Or maybe it was a second bird. Time and again, she'd return from the field and study the images from her camera traps. Time and again, she realized the birds had been a footstep or two away from her. She still had so many questions: did they migrate or stay in South Carolina? If they left, where did they go? Florida? "The biggest thing I've learned is how much I still don't know about them."

And there are other questions, including ones that fly beyond her field cameras and data: how do we protect something we love?

Hold it close so it's all but hidden?

Or share it so others love and protect it, too?

---

Yes, Christy Hand will talk about the bird. Science isn't something you hold back. And we could learn a lot from this bird and its plight. The stakes go beyond black rails. Since 1970, North America has lost nearly 3 billion birds—one in four, researchers reported in *Science* magazine. The National Audubon Society calculated that if human beings don't stop burning fossil fuels and find other ways to slow climate change, two-thirds of North America's birds will be extinct in eighty years. The climate will change so fast that many birds won't adapt in time, especially little black birds that nest on a narrow edge of habitat.

So, during her next research trip into the field, she'll take me to that edge—and perhaps the rails.

We walk into the marsh with a parabolic dish. It's about the size of a small umbrella. If she hears a call, she'll point the dish toward the noise to get better audio. We probably won't hear anything, she says. Then again, the bird songs were loud a few weeks ago on a perfect night when the sun was setting and the marsh was ablaze with life. But they seemed to be quieting since then.

Dragonflies buzz, and a light breeze sends ripples of shadows across the grass. Spiders make silk between a few blades, and that honey light casts everything in deeper shades. Christy puts on her headphones. She plays a soft *kickee-doo*. She doesn't want to play too many recordings, doesn't want to confuse the rails, add stress to the birds' lives. If they're here.

And then ...

Over there, about fifteen feet away in the grass. *Kickee-doo!*

And behind me. *Kickee-doo!*

And, over there, toward the sun. *Kickee-doo!*

They're here, several adult black rails, though still unseen, like ghosts. And then, through her headphones, Christy hears something else, and the world vanishes: the pandemic, the uncertainty stoked by a rapidly warming planet. And in its place is only the quiet day-end exhalation of the marsh—and the bubbling and hopeful sounds of at least two newly hatched black rail chicks.

The Santee Delta's old rice fields are still visible.
Photo by Lauren Petracca of *The Post and Courier*.

# OUR SECRET DELTA

THE SOUTH CAROLINA LOWCOUNTRY IS A LAND OF MANY EDGES, some obvious, others hidden. The tides blur things. Because of our low elevation and the moon's pull, vast areas of land and water trade places twice a day. This makes our edges spongy instead of hard. In this soppy zone, green and gold strands of Spartina grass poke from mud so black and gooey it resembles tar, and the land's relationship with water is so intimate, saltwater sometimes pours from the ground like sweat.

I was thinking about these changeable edges when Glenn Smith, the newspaper's projects editor, suggested we take a deeper look at the Santee Delta. My knowledge about the delta was thin, even though it's just an hour's drive north of Charleston. I knew it mainly as a beautiful expanse you saw on the Highway 17 bridge over the Santee. In the coming months, though, we'd learn how it was so much more. How it was a storied place with secrets hidden by neglect and time, a place that's difficult to love, especially when deer flies hit your skin like hailstones. We'd learn how the delta is at once one of the least trodden and most vulnerable places in the Southeast, and that a rapidly warming climate could change it forever. And that all of this takes time to understand. And a boat.

But if you did take the time, and if you found a boat and loaded up on bug repellent, you discovered how the delta opened like an old book, one about the ebb and flow of money, power, and wont. And water.

Always water, whether the stories are new or old, tragedies or mysteries.

But we're getting ahead of ourselves. So, before we get to the stories and secrets, let's get our bearings, let's follow the water.

---

From the Blue Ridge Mountains of North Carolina, water flows toward South Carolina. It gains force as foothills give way to South Carolina's Midlands, forming the Saluda, Catawba, and Broad rivers, brown with loam.

Past Columbia, the land flattens and the rivers expand like lungs into the great cypress swamps of the Congaree. The water slows at Santee Cooper's dams, forming the shallow lakes Moultrie and Marion. Then it shoots through spillways. Some goes into the Cooper River toward Charleston's busy harbor. But most of it pours into the much-less-busy Santee.

With no real slopes to guide it now, the Santee meanders like a haphazardly thrown rope until it splits in two, the North and South Santee. Finally, in the flats between Georgetown and McClellanville, those strands meet an opposing force, the rising tides of the Atlantic.

This meeting place is where the delta becomes the delta, where sediment flowing from Upstate fans out and makes marshes and barrier islands. And, because this land is low, and because of the tides, the ground isn't always solid. It's something in between: swamps so quiet you hear blood rush in your ears, old rice fields with flocks of blackbirds that lift as one and move back and forth, like a conductor's hand.

All told, the Santee Delta drains an area the combined size of Massachusetts, Vermont, and New Hampshire. Its floodplain below the Santee Cooper dams fills 550,000 acres, an area seven times larger than the city of Charleston, and most of it is undeveloped. The delta's marshes and swamps are home to more than one hundred threatened or endangered plant and animal species. At least one plant grows nowhere else on Earth.

Despite its size and uniqueness, the Santee Delta has long flowed in the shadows of the more celebrated and smaller ACE Basin estuary near Beaufort and the Santee's immediate neighbors, the Francis Marion National Forest and Cape Romain National Wildlife Refuge. We needed a guide to help us uncover the delta's secrets, and we soon found one in Richard Porcher, who one late afternoon found himself lost in a place he knew better than almost anyone else.

---

Richard Porcher is a sturdy man of eighty years, with a trim frame, white hair, and a gleam in his eyes, especially when he talks about the Santee Delta's mysteries. He has an increasingly rare Charleston accent. He deploys this accent in baritone bursts of stories and facts. Friends call him "Hoot Owl."

His Carolina roots stretch back twelve generations to 1685. That's when his ancestors arrived—French Protestants called Huguenots, which is why his name today is still pronounced "pour-shay." He's quick to say that his forebearers fled France because of religious persecution, but once here, the bullied became bullies and worse, importing people in bondage to work in the fields. On this afternoon, Richard looked like a colonial explorer with tall rubber boots over blue corduroy pants and a canteen slung over his shoulder. His ropy arms protruded from his weathered tan shirt, his skin showing signs of age and cuts from journeys in the field. He carried a large wooden staff. "For the snakes," he half joked.

Some of his ancestors were acclaimed botanists, and he'd long felt the calling to continue in their boot steps. He began exploring the Santee Delta forty years ago when he taught biology at The Citadel, South Carolina's military college. He has since written definitive books about the state's wildflowers and sea island cotton. But the Santee Delta's rich human history also captured his imagination, especially its emergence as the center of the South's "Rice Kingdom," so he wrote a definitive book about rice.

As part of that research, he mapped every rice-related structure he could find in the delta: rice mill chimneys, rice field shelters, remnants of slave quarters. This was no easy task. The delta is vast and has few roads. Rivers and creeks are the only routes to some spots. And you can't see far from the low vantage of a boat seat. So, he developed a ruin-hunting technique: from your perch in a boat, keep your eye trained for trees and bushes. That meant high ground. And high ground meant better odds that people once lived there. Then get out.

"You would have no idea what's here if you just stayed in your boat."

---

Earlier in the afternoon, Richard Porcher had climbed into a boat with Chris Crolley. When Chris was in his twenties, he carried around Richard's *Wildflowers of South Carolina* until it was worn out. Now, years later, Chris owns Coastal Expeditions, one of the few outfitters that take people into the delta.

"Sometimes I wonder how come everybody doesn't know about this place," Chris said. The delta's stories help explain South Carolina's complicated past and possibly the shape of things to come. "There are secrets here that people should know about, mysteries. And the clues are still here." One of those clues was downriver: the ruins of a former slave village on Crow Island.

Richard and Chris had begun their search from Pole Yard Landing, a public boat launch on the North Santee. As they made their way downriver, they spotted stands of cow-itch, a plant also known as orange trumpet-flower. A black skimmer bird sailed by, its long beak just inches from the water's surface. Bald cypresses lined the bank, and Chris noted how their intertwined roots give the trees stability in soft mud. At low tide, the roots were exposed. "You can see how they hold hands," he said.

Before the Huguenots and British settled the Lowcountry, the Santee Delta was a vast forest of bald cypress and tupelo. So much freshwater poured from the mountains and Midlands that the river kept the saltwater at bay. As a result, the cypress forests grew surprisingly close to the beaches. The evidence was still here in the form of a hulking gray cypress stump. "That one there is older than Jesus," Chris said.

Richard Porcher studied a copy of a map drawn in 1875, glancing up now and then, looking for trees. "There it is," he said. Crow Island.

Chris piloted the boat toward a mud bank. An alligator drifted in the murky water nearby. Fiddler crabs skittered toward shelter as Richard stepped off the boat into the muck. His boots made a sound somewhere between a crunch and a squelch. He filled his nostrils with the smell of wet mud and brine. He pulled himself up a slippery embankment and into grass two feet above his head. Then, instead of planting his stick to walk, he turned it sideways and used it as a plow, pushing down cordgrass to blaze a trail.

Slave settlements such as the one on Crow Island once were scattered across the delta. This one likely had eight or nine cabins that housed as many as sixty people in bondage. These villages were different from the alleys of cabins next to plantation homes you might find around Charleston. Slaves working the rice fields were much more isolated.

Richard Porcher first found the Crow Island ruins in the early 1980s and felt as if he'd discovered a lost city. No one had documented its existence. No one had done excavations to better understand the lives of those who had been enslaved there.

Hunting for the ruins once again, Richard pushed through the grass and branches of scrubby trees. He stopped and pivoted. He wiped his brow with a handkerchief; his map fell from his pants pocket. He walked deeper into the grass. The blades sliced his arm; his left elbow bled. "I'm a little discombobulated." Dragonflies and bees buzzed nearby. He took a swig from his canteen and confessed: "I don't know where we are."

---

The Lowcountry is well-named, and the Santee River merges slowly with the Atlantic. As river and ocean meet, the lighter freshwater forms a layer that rides over the heavier saltwater from the sea. Like clockwork, the tides lift this freshwater layer several feet high. This lifting phenomenon triggered an agricultural gold rush that forever changed the delta.

In the early 1700s, planters chopped down the delta's great cypress forests. Enslaved workers with hand tools and oxen built enormous rectangular dike fields. Then they harnessed those tidal layers, installing wooden gates called trunks at certain heights along the dikes.

At high tide, you opened the gates so the freshwater layer flooded the fields.

At low tide, you drained the fields and harvested the rice.

With the dikes and trunks, rice growers suddenly could flood and drain fields without worrying about droughts. It made rice cultivation more predictable and profitable, and spawned even more dike work across the Lowcountry.

In just one plantation along the Cooper River, enslaved workers built fifty-five miles of dikes, moving earth equivalent to three Egyptian pyramids. The swampy and wide Santee Delta was even better suited for tidal control. And, by the American Revolution, the delta's rice fields stretched like a handmade quilt toward the horizon. The rice was called "Carolina Gold," and worldwide demand triggered a rush that turned the Lowcountry into a landscape of misery and money. Planters built great houses on bluffs overlooking the dikes; they built mansions in downtown Charleston. Away from the white columns, they established slave settlements in the swamps, such as the one on Crow Island and another just upriver on Tranquility Island.

It was a miserable place to live and work. Standing water bred clouds of mosquitoes. A British observer wrote in 1775 that slaves stood, "ankle, and even mid-leg deep in water which floats in oozy mud; and exposed to a burning sun which makes the very air hotter than the human blood; these poor wretches are then in a furnace of stinking, putrid effluvia."

Hurricanes came without warning, and one in 1822 hit the Rice Kingdom like a sucker punch. Nearly every house in Charleston and Georgetown was damaged or destroyed. Slave villages and plantation homes alike were swept away. Survivors were found drifting on pieces of lumber. Hundreds of enslaved people died. On Murphy Island, near the mouth of the delta, fifty people in bondage drowned or were crushed by falling debris. The headline in *The Charleston Courier* said simply: "Dreadful Hurricane!" Malaria, yellow fever and dysentery were rampant. In these harsh and exposed conditions, many enslaved workers survived just a few years, so planters sought replacements, paying more for those with rice-growing expertise in West Africa. "Gold Coast or Gambias are best," Henry Laurens, a Charleston planter with Huguenot roots, wrote in 1755.

Rice cultivation made Laurens and other Lowcountry planters rich. The Lowcountry exported sixty million pounds of rice a year. And the sweat-soaked Santee Delta was the Rice Kingdom's beating heart—a kingdom that would become even more profitable and entrenched after an English inventor's rough arrival.

---

Not too far from where Richard Porcher pushed through the cordgrass on Crow Island, a storm in 1786 blew a ship ashore. On board was a British inventor named Jonathan Lucas, headed to Charleston to make a new life.

Lucas was thirty-three and came from a family of millwrights. After the shipwreck, Lucas noticed the agonizing way slaves separated hulls from the grain: using heavy wooden pestles, slaves pounded like pistons on the grains, hour after hour. It was tedious and back-breaking work. Enslaved workers eked out a few pounds of rice per day. Lucas had an idea. Combining existing technologies, he designed new mills that for the first time used huge millstones to remove the hulls. These millstones replaced the pounding work done by enslaved people and their mortars and pestles. Lucas also harnessed tidal currents to turn the millstone and power a system of buckets, pulleys, and wind fans that separated and polished the grains, which poured down chutes, ready to ship.

His automated mills revolutionized rice production. Just three people could run a Lucas mill, allowing planters to deploy more enslaved workers to their fields. Landowners across the Rice Kingdom were eager to cash in and sought Lucas out. Within a few years, he'd built fifteen water-powered mills across the Lowcountry. Lucas's contributions to agriculture were akin to Eli Whitney's cotton gin, also invented in the 1790s. They were "monumental and ingenious," Richard wrote in his book *The Market Preparation of Carolina Rice*. But while Whitney's cotton gin is cited in schoolbooks, the Lucas rice mill faded into obscurity.

Yet the evidence of Lucas's impact is as real as the mansions in Charleston that face Fort Sumter. Like Eli Whitney's cotton gin, the Lucas rice mill increased demand for more enslaved people, binding the South's fate to agriculture and injustice. By 1850, Joshua John Ward of Georgetown County alone owned eleven hundred people in bondage. (His land includes what today is Brookgreen Gardens.) Rice planters such as Ward grew richer by the year, as the South moved ever closer to a collision with the industrializing North.

---

On Crow Island, Richard Porcher pushed deeper into the cordgrass, still hunting for the ruins. By now, the late afternoon light was the color of honey. He walked into a clearing. As if someone had opened a door, a lake of grass spread out before him, a remnant of an old rice field. The sense of isolation was as fresh as the wind gusts that run along the grass.

Four decades ago, on his first trip to Crow Island, Richard tried to imagine what it was like for those who lived here. The questions spun in his mind: what did they talk about at night? What kind of aspirations did they have? What did the numbing work in the fields do to them and their children?

Like Richard Porcher, Vennie Deas Moore also wondered about these forgotten lives. It was personal for her. Her descendants once toiled at Hampton Plantation, now a state historic site, and possibly the rice fields of Tranquility Island, just upriver from Crow. As a cultural historian, she spent years studying the lives of African Americans in the delta, telling their stories in books, on film, and in oral histories. Then, one day a few years ago, she traveled with Richard to see Tranquility Island for herself.

It was thick with flies and cordgrass, she later recalled. She felt the island's isolation. Were there snakes about? Wild boars? She didn't know. One thing was clear, the mosquitoes were vicious. She thought about her ancestors working in the unforgiving sun, day after day, as mosquitoes and flies feasted on them. Anything but tranquil. Then she took a step and suddenly was up to her waist in pluff mud and sinking fast. She yelled for Richard to help. He rushed to her aid with a switch of thick grass. Grab on, he told her. And don't move. You'll only sink more. Using the grass, Richard pulled her to safety. She was covered with a crust of sour-smelling goop.

Later that night, Vennie Moore told her mom what happened. Her mother gasped.

"Don't you ever go back out there again," she scolded Vennie. "Do you know we lost whole cows out there?" Vennie, now seventy years old, pondered the legacy her ancestors left behind, the sprawling dike work, the courage and determination it took to survive, the riches they created for others. Mansions and monuments to this wealth still stand in downtown Charleston's old streets. But she knows that deep in the delta cordgrass, time and rising seas are erasing the evidence of this taking.

---

As the Civil War loomed, Lowcountry rice planters reaped more riches, producing 119 million pounds of rice the year before the first shots on Fort Sumter. But after the war, the newly freed left the fields. Dikes failed; weeds choked ditches; Europeans imported cheaper rice from Asia; more hurricanes swept through. Rice farmers in Texas and Louisiana began using mechanical threshers. But these machines sank in the Lowcountry's soupy soils. The Carolina Gold rush was over, and the Santee Delta began to empty its people.

Taking their places were clouds of ducks, millions of migrating waterfowl—and hunters soon followed, including President Grover Cleveland. On a trip in 1894, Cleveland got stuck up to his thighs in the marsh muck. A local African American guide wrapped his arms around all 260 pounds of the president and wrestled him out of his hip boots and onto solid ground. Cleveland took a swig of whiskey and laughed it off. But newspapers told and retold stories of Cleveland's hunting exploits, stories that also revealed how former rice plantations were being sold at dirt-cheap prices.

Word spread about skies full of ducks, the cheap land, and soon trainloads of hunters left New York for South Carolina when the weather turned cool. Hunting clubs formed to cater to these new well-heeled hunters, including the Santee Club, created in 1898, and later known as the Santee Gun Club. Members of the Santee

Gun Club were among the nation's wealthiest people: oil company presidents, telegraph moguls, department store magnates, du Ponts, a Rockefeller. By 1900, the Baltimore American newspaper described the club as "the most influential gunning club in the United States."

The Santee Gun Club would eventually encompass twelve former plantations and more than twenty-four thousand acres. More wealthy industrialists arrived, snatching up other nearby plantations. This new land rush generated some tension. A Charleston *News & Courier* reporter wrote about the Northerners arriving in private train coaches, "bringing with them white servants who turned up their noses at everything they saw around the countryside" along with their "polo ponies, snooty foxhounds ... imported whiskeys, antique furniture, and sporting clothes fashioned in London that never seemed to fit." Locals called the land rush "the second Yankee invasion."

Newcomers included William Yawkey, a lumber and mining millionaire who acquired large tracts near the mouth of the delta. Yawkey also owned the Detroit Tigers. Baseball legend Ty Cobb spent time at Yawkey's Santee plantation, practicing on fields next to giant live oaks. Yawkey's son, Tom, later learned to hunt there, and when he inherited the land, he reportedly said: "I hope I can do some good with it." Tom Yawkey eventually bought his own baseball team, the Boston Red Sox, and photographs of Red Sox great Ted Williams still hang in the "Playroom," a stately wood-paneled building where Yawkey and his wife received guests. The Yawkeys lived in more modest digs nearby—a mobile home. Yawkey, an introvert, joked that they lived in a trailer, so guests had nowhere to stay.

Over the decades, farmers and developers and dam builders drained wetlands across the Carolinas and the country. As wildlife habitat disappeared, so did the ducks, and Yawkey grew concerned. He'd seen declines in the delta, especially after the utility Santee Cooper built the dams for Lakes Moultrie and Marion. He stopped hunting and started managing his land as a refuge. He opened it to researchers, including those who studied alligators.

Like human beings, alligators have long life spans and accumulate pesticides and other industrial chemicals in their tissues. Long-term studies on alligators and pollution also could tell us things about ourselves. In time, researchers knew the gators by name: Big Bertha was the oldest and largest female—nearly ten feet long. They called others Bette Davis Eyes, Truck Biter, and Grover. As years passed, this work grew more valuable—precious because there were so few large coastal areas that had been set aside for wildlife for so long.

Then, in the mid-1970s, two events converged like rivers: first, in 1974, the Santee Gun Club donated its massive holdings to The Nature Conservancy, and the conservation group deeded most of the land to the South Carolina Department of Natural Resources. An editorial in *The New York Times* called it "one of the most valuable single gifts made in the interests of American conservation—almost comparable to the Rockefeller gifts of entire national parks."

And in 1976, Tom Yawkey died. In his will, he left 20,000 acres to South Carolina. Yawkey also bankrolled a nonprofit foundation to manage the land as a wildlife preserve, a boon to taxpayers who suddenly had a large new protected space but didn't have to pay for its upkeep. Soon, private plantation owners began donating conservation easements to The Nature Conservancy, Ducks Unlimited, and Lowcountry Land Trust. Together, this combination of public and privately protected land formed a massive preserve, one sandwiched between two of the fastest-growing metropolitan areas in the country, Charleston to the south and Myrtle Beach to the north. Today, on guided visits at the Tom Yawkey Center, you can see a flock of pink roseate spoonbills explode from the marsh in flight, far from their normal haunts in the Everglades. You might see the Carolina hedge-nettle, which grows only in one spot in the Yawkey refuge. The federal government calls the plant "critically imperiled."

And you might stand on one of the center's dikes, and with that small bit of altitude, look onto a prairie of luminescent yellow and green that's so expansive it seems to end only when it meets sky.

---

The sky, like the climate, is ever changing. But the delta in the past four years has experienced unusual motion, a faster pace, as if its chapters had been shortened.

Year after year, skies turned heavy and gray. Clouds moved into the Lowcountry like invading forces and then unleashed massive amounts of water and disruption: the "thousand-year" storm of October 2015; Hurricane Matthew in 2016; Irma in 2017; Florence in 2018; Hurricane Dorian in 2019; and Hurricane Ian in 2023. The storms felled the delta's trees, and the ocean ate into its barrier beaches. Because it's flat and porous and mostly unpaved, the delta absorbed these storms like a sponge, limiting the damage. But rising seas could, in a few generations, reshape the landscape for good.

In the Tom Yawkey Wildlife Center, rising seas have eroded riverbanks and made it more difficult to drain ponds. Saltwater is finding its way farther upriver, winning its long tug-of-war with the freshwater coming from the mountains. Compounding the problem, Santee Cooper's dams have bottled massive volumes of silt that once helped build Cedar and Murphy islands. Starved of beach-building sediment, they're quickly losing ground. On Murphy Island, especially, an important African American historical site is under the gun. After the devastating hurricane in 1822, planters built circular brick towers in the delta for enslaved workers to ride out storms. "There's nothing like them anywhere else," said Brent Fortenberry, a Texas A&M University architecture and preservation professor. The Santee's planters didn't build them necessarily out of compassion, he told me. "They did it to protect their labor force. But they're an indelible piece of the Santee Delta puzzle and South Carolina's past." And yet, on a recent visit, the storm tower looked as if it was about to be swallowed by rising seas on one side and a jungle on the other. Brick thieves have removed whole sections. A dying tree grew in the middle, poised to fall in a strong wind.

Richard Porcher surveys Crow Island deep in the Santee Delta.
Photo by Lauren Petracca of *The Post and Courier*.

Over on Crow Island, Richard Porcher continued hunting for those slave ruins, still lost. Time to backtrack. With his sideways walking stick, he marched toward the river where Chris Crolley waited in the boat. Richard climbed in, sat for a few minutes and took another swig from his canteen. "I was lost, and I don't mind admitting it," he said.

Chris smiled. They both understood how wild the Santee Delta is, even though human hands have shaped it for centuries—and how rare it is to find places so undeveloped that you can so easily lose your way.

Richard wasn't about to give up. He's always had a motor, but at eighty he has the urgent rev of someone who knows his journey will end sooner than later. He and Chris formed a new plan: head upriver toward Tranquility and then float back to an inlet near Crow.

They landed near the inlet. With his bearings reset, Richard stepped back onto the old dike, mowing a new path with his walking stick. Suddenly, he found a clearing. The grass was shorter here and opened onto a savanna that once grew rice. Here it was: piles of old bricks from fireplaces and chimneys, old bricks that warmed the enslaved people who worked in those dike fields. A thick cypress post, hand-hewn into a beam, was planted into the soil nearby. "There's so much we don't know about this place."

Piled there in the grass, the bricks weren't much to look at. But around him, everything felt open, the eroding old rice fields, the waves of cordgrass, the quiet. It was enough to get his mind spinning again: what was it like at night, when the stars came out? What was it like in August when the heat never died? What was it like for an enslaved mother to tend to her sick child? Where did they bury their dead? He'd seen no evidence of graveyards. Did they ever go to the mainland? Or were they confined their whole lives to the delta? "The cultural history, the slave history, how it changed, everything fits together out here," he said.

Those stories are easily forgotten. "History closed its pages on the Santee Delta."

But professors like Richard Porcher know that books are designed to be opened again and again. And the Santee Delta—so manipulated by human hands yet so wild despite it—is no different. Richard turned back toward the river, a smile on his face, energized by the rediscovery of Crow Island, the sun sinking lower as he plowed through the last stand of cordgrass, now turning a richer shade of green in the gathering twilight.

Dunes in Senegal's Lompoul Desert near the border of Mauritania.
Photo by Andrew Whitaker of *The Post and Courier*.

# THE SAHARA CONNECTION

WE KNOW THAT AS THE PLANET SPINS, WINDS NATURALLY SLANT toward the equator, creating the trade winds.

We also know that tropical storms form because the sun shines bright where these trade winds blow, turning seawater into sky-high clouds of steam that inevitably collapse in spasms of thunder and rain.

And we know from history and physics that a few of these superheated air masses will form in the ocean off West Africa and spin counterclockwise, slowly at first, then faster until they gather enough momentum to flatten cities and, if hooked into some fantastic electric grid, pack enough energy to light every bulb on this rotating blue orb.

Beyond these certainties, hurricanes challenge us with questions. Where will they go? How many will we have this season? Will my insurance rates go up? Should I get a backup generator? Like the humidity, anxiety settles in for the summer.

I'd covered more than seventeen hurricanes for the newspaper, sometimes watching the area off West Africa launch storms like cannonballs toward the Caribbean and Southeastern United States.

One of those cannonballs was Hurricane Hugo in 1989, which shot across the Atlantic and plunged into South Carolina with such force that people in Charleston began measuring time differently: Before Hugo. After Hugo.

The ocean off West Africa is the hurricane nursery where our worst hurricanes were born, and this origin often reminded me of another important steering current. Slavery and its legacy had long shaped South Carolina's past and present, and watching the trade winds carry those tempests toward us felt like karmic revenge.

I also knew that the trade winds carried something lighter—dust from the Sahara Desert. I'd read years before how the sun heats the desert's sand, and that the rising air pumps the dust into the trade winds and sometimes created orange films on car windshields in Georgia and made Lowcountry sunsets even more beautiful. How could particles from an African desert fly so far? The phenomenon was another reminder of our connectivity with Africa.

Then I learned about something that made me even more interested in the Sahara—how clouds of dust over the Atlantic can snuff out potential hurricanes like blankets on a smoldering fire. Was Saharan dust a good thing for the Lowcountry? Apparently so, at least when it came to hurricanes and sunsets. But what was climate change doing to these shifting sands in the sky?

My mind drifted toward the source, West Africa. That's where I wanted to go. I'd talk to scientists who had seen the towering clouds of dust consume their cities, felt the grit in their food. Where finding answers was personal.

I learned about a scientist at Penn State University named Gregory Jenkins, who set up a network of climate scientists in West Africa. I called Gregory, and he mentioned that some of his colleagues in Senegal had done groundbreaking climate research, discoveries that had been downplayed in part because of science's long tilt toward Europe and North America.

Searching for dust and answers, I flew to Senegal in May, near the end of the dust season. From a storytelling standpoint, I hoped to see a storm called a *haboob*, Arabic for "blowing furiously." I'd seen ominous photos of haboobs on the Internet that showed curtains of dust approaching towns like tidal waves.

But on the long taxi ride into the bustling capital of Dakar, I noticed something missing. The air seemed to have no dust at all.

Moussa Gueye surveys the beach where waves destroyed a family member's home when he was a child. Photo by Andrew Whitaker of *The Post and Courier*.

Gregory Jenkins had urged me to talk to one Senegalese scientist in particular, Moussa Gueye. But Moussa wasn't in Dakar; he taught applied mathematics at USSEIN, a university in Kaolack, a city of about three hundred thousand people three or four hours away. Great, I thought, maybe we'll see a haboob on the way.

With me was Andrew Whitaker, a photographer from *The Post and Courier*, and Borso Tall, a local journalist who would help us with logistics and translations. Borso also had surprised us with the news that she had gotten married a few days before our arrival. I looked at Whitaker and then her. "This has to be the worst honeymoon ever," I said, feeling a little guilty for our trip's timing.

"No worries," she said, adding that her new husband, Pape, was traveling in another car behind us, another surprise. And she was happy to be in Kaolack, where she'd grown up. She told us the city was known for its peanuts, and we made a pit stop at a processing plant, where peanut piles the size of sand dunes rose next to scattered groves of thorny acacia trees. Moussa Gaye's office at USSEIN was a short drive from there.

Moussa Gueye was a tall and thin man with a scruff of hair on his chin and an easy smile. Like all good math teachers, he's good at making abstract math concepts relevant to his students, which he sometimes does by offering his life as a non-mathematical proof.

He told me he was born about seventy miles west of Kaolack, in a town on Senegal's coast called Mbour. His father and uncle were fishermen who plied the waves in wooden pirogues, long banana-shaped boats often painted green, yellow or red, the colors of the Senegalese flag. The beaches in Mbour had plenty of space in the 1990s for him and his friends to play soccer, with room to spare for the many pirogues that lined the shore. But the waves seemed to grow closer year by year.

Then, one night when he was six years old, the Atlantic poured in. Like thieves, the waves ransacked his uncle's house. Families fled for their lives as the sea stole their belongings. As the water receded, Gueye tried to make sense of what had happened. No evil person was responsible. It was something more powerful. His young mind began to spin with questions about the forces of nature and how they work.

His father had seen how the fishing grounds were being depleted, and he urged his son to focus on school instead of following him into the sea. Gueye eventually graduated from Cheikh Anta Diop University in Dakar and traveled to France for his doctorate. There, he had another pivotal moment during a conversation with an advisor about a global climate model.

Climate models are humanity's best estimate of what might happen to the Earth's atmosphere and oceans in the future. They work by breaking the planet's atmosphere and oceans into cubes. Each cube is packed with data on temperature, ice, and winds. Computer programs then analyze what happens in those cubes to identify patterns. Yet, these models depend on real-world measurements in those cubes. Without accurate data, they spit out inaccurate forecasts. There, in France, Gueye realized that the model didn't properly account for one of the most important migrations in the world—the Sahara Desert and its dust.

This was a glaring failure, Gueye thought at the time. Even as a young child, he'd seen how dust clouds made it harder on everyone to breathe. And the Sahara was such an enormous presence on the planet, a desert as large as the forty-eight continental United States. But he also knew that this massive region had few long-term weather sensors, which made the Sahara a desert of data. The question nagged at him. The grand movement of the Sahara's dust—how could it not be part of the larger climate equation?

---

Here it's helpful to look at the history of dust research, which won't take much time because it's a surprisingly young field.

In 1966, Joseph Prospero was at the University of Miami and studying the effects of seawater bubbles and what happens when they burst. By chance, he met a group of British researchers who had been searching in Barbados for evidence of cosmic dust. They had set up filters that captured air from the trade winds. But instead of particles from space, they found their filters caked in red dust. Prospero took over the Barbados filtering project, documenting for the first time the seasonal pulse of Saharan dust across the Atlantic.

Next, Prospero and Toby Carlson, then with a federal National Hurricane Research Laboratory in Miami, pinpointed a river of dust over the Atlantic they dubbed the "Saharan Air Layer." The layer typically begins five thousand feet above the ocean and rises another ten thousand to fifteen thousand feet. For these and other discoveries, Prospero is called the "Father of Dust."

Scientists know now that every year on average, the Sahara launches 182 million tons of dust into the atmosphere. Some of the largest dust sources are in Mauritania and Mali, where particles have an orange tint, and the Bodélé Depression in Chad, which has a whitish tinge. You can easily see the Bodélé Depression on satellite images: a white eye-shaped indentation the size of South Carolina.

Depending on the winds, African dust may drift north to Europe. During winter, dust sometimes coats ski slopes in France and Spain with a terra cotta-colored film that makes them look like sand dunes. Dust from Chad has been found as far north as Greenland. When dust and rain mix, you end up with a rainstorm of muck. In 2022, a dust cloud from the Sahara passed over Paris, turning the sky the color of pumpkin. Then it fell in the United Kingdom. Forecasters called it a "blood rain."

But much of Africa's dust flows west, toward the trade winds—so much dust that if packed into semitruck trailers would fill seven hundred thousand of them. This sky-borne convoy then crosses the Atlantic. When dust arrives in the Caribbean, emergency room visits increase and asthma complaints spike. Dust is rich in iron and phosphorus, so when it lands in water and forests, it fertilizes phytoplankton, those microscopic jewels that produce half of the oxygen we breathe. Dust also carries harmful microbes. Away from the Sahara, researchers in China found spores from Candida fungi on dust from farmlands, dust that caused heart damage in children. In Senegal, researchers found neurotoxins on dust particles from the Sahara, the same toxins you find in molds that trigger allergies.

In June 2020, a plume of dust flowed off Western Africa and floated five thousand miles into the Gulf of Mexico. It was a layer of dust the size of the United States and two miles thick, and it was among the largest dust clouds NASA had seen in fifty years. It gave the daytime sky in Texas a sickly white glare and the sunsets a burnt orange hue. Its tiny particles triggered allergies in Louisiana. It made headlines as it moved as far north as Illinois. Meteorologists nicknamed it Godzilla.

Prospero and his colleagues had opened a door into a whole new arena of science, one that captured the imagination of Gregory Jenkins, the scientist at Penn State. Jenkins also felt the Sahara's pull, wondering how the Saharan Air Layer affected hurricanes that strike the United States.

West Africa would be his most important teacher.

Like Moussa Gueye in Senegal, Gregory Jenkins also grew up wondering about the forces of nature. The clouds, snow, stars—questions popped into his head as soon as he walked out the door.

His family lived in a row house in Black Bottom, a close-knit, low-income neighborhood in west Philadelphia. His father, Kirby Jenkins, was born in Walterboro, an hour west of Charleston, and later moved to Philadelphia. His father fought in World War II's Battle of the Bulge and worked at the Philadelphia Shipyard after the war. But he died of a brain hemorrhage when Gregory was a child. His mother, a beautician, died when he was a teenager. The deaths left him off balance for a time as a young man, balance he'd regain as his science career spun forward.

Balance, that was what he'd eventually look for in the rotating planet's winds and seas. He could see the rebalancing in how polar air moved south and warmer air moved north, how ocean currents moved water around the planet like a radiator, how dry air from the Sahara met moist air from the tropics, how a rapidly warming climate upset this balance, especially in Africa.

In the early 1990s, Jenkins went to Niger and Senegal after he finished his doctoral degree at the University of Michigan, and his curiosity gained even more momentum. Countries in and near the Sahara had suffered through a catastrophic drought that began in 1968 and continued through the 1980s. It was the worst drought to hit the planet in the 20th century, one eventually linked in part to air pollution in Europe and North America and rapidly warming water in the Indian Ocean. The Sahara launched higher-than-normal levels of dust during these two decades, as studies from Prospero's filters in Barbados would later show. The drought and high dust levels also coincided with a two-decade lull in hurricane activity.

In Senegal, Jenkins saw the orange dust, the poverty balanced by social ties, and realized he'd found the place he'd study for the rest of his life. Jenkins visited Senegal year after year, lugging weather instruments with him, his affection for West Africa growing, as well as his questions: how did the drought affect rainfall patterns and dust? How did this dust affect hurricanes?

"But our tools for this region weren't good," Jenkins recalled. They had satellite information, but that wasn't enough. They needed real measurements—temperature, humidity, wind speeds. They needed to set their feet on those dunes and dust sources, capture what was happening in the skies above these poorly measured places. "The United States is covered with sensors, but there are just a few in West Africa for 350 million people," Jenkins said. "How can you have this kind of imbalance?"

In the early 2000s, he returned to Senegal for eight months, working with scientists there to collect key climate data across West Africa. Researchers tracked movements of dust; they monitored the monsoons; more data on dust flowed into climate models, improving their precision. Then more West African researchers published studies in international journals, including Moussa Gueye, who spent five years crafting a fix for the French climate model, gathering data and coding programs to simulate the migration of Saharan dust. "The voices of scientists in West Africa needed to be heard," Jenkins said.

Today, from Dakar to Saint-Louis to Kaolack, Senegalese scientists have a deep understanding of how their dust-laden storms connect with ours. They know this connection begins with heat, the blistering sun that bakes the Sahara in Mali, Algeria, Chad, and Mauritania. As the air rises from that heat, winds from the north rush in to fill that imbalance.

When the wind has enough velocity, it scoops up the dust. During the dry season, these winds are called harmattans, a West African word derived from the words "to blow" and "tallow," animal fat used to keep skin from drying out. When the harmattans reach Senegal, they paint the sky a smoky brown for months.

This dry season typically ends in May and June. Just before the rains come, storms churn through the Sahara, hoisting enormous amounts of dust. These are the haboobs, Arabic for "strong winds." As haboobs cross Senegal, they create mile-high walls of brown. Visibility can drop to almost zero. The haboobs can be scary, but they often precede the season's first rains. So, fear of the storm is balanced by relief that the dry season is over.

There's a second factor at work—the jet stream flowing across Africa. Dust in this westward-moving air current eventually meets the rising winds off the Senegalese and Mauritanian coasts. Like a springboard, the collision of these winds pushes dust even higher. This is where the Saharan Air Layer takes its shape over the Atlantic. The Saharan Air Layer typically has half of the moisture of the air below. This warm and dry air makes it more difficult for clouds to punch through, smothering potential storms.

If the layer is especially thick with dust, it also blocks some of the sun's energy. Less sun means less heat going into the ocean, less fuel for developing cyclones. In other words, the coastal zone off western Africa is often where storms are born or die.

---

I asked Moussa Gueye and his colleagues whether climate change had affected these forces. All said they'd seen the changes firsthand and in their data: more punishing deluges, more frequent droughts, more severe dust events. Disruptions that forced farmers to move to new megacities such as Dakar or cram into pirogues headed toward Europe, ripples of social and atmospheric change that spread far beyond the Sahara. You can't understand what's happening to the world, Gueye said, unless you understand what's happening in Africa.

"Will there be more dust or less dust?" I asked, thinking about how more dust would be a good thing for Charleston—fewer hurricanes—but not so good for my new friends in Senegal who would have to breathe it.

"It's not a simple answer," Moussa Gueye said.

Scientists at NASA had recently predicted that a warming atmosphere would weaken the trade winds, and that a weaker African jet would allow bands of thunderstorms in the tropics to migrate north into the Sahara and its edges.

This would dampen the dust as the climate tries to rebalance. Less dust in the air meant the ocean would absorb even more of the sun's heat, creating a feedback loop that accelerates the overall warming trend. The models predicted a 30 percent decrease in the amount of dust over the next twenty to fifty years. Less dust meant fewer nutrients would enter the oceans and fertilize our oxygen-producing phytoplankton. Less dust to snuff out our hurricanes.

At the same time, Gregory Jenkins and Moussa Gueye saw evidence in the models that pointed to a different horizon—that a warming planet will make the Sahara even hotter, generating bigger blasts of Saharan dust over the next thirty years. "But the truth is, we don't know for sure," Gregory Jenkins told me. "That's why we need more data."

It's this uncertainty that fuels their quest for answers. It's why one morning in Senegal, Moussa Gaye wanted to show me the place that first incubated that search.

After teaching a class, we drove out of Kaolack on a hot and mostly dust-free two-lane road. Along the way, we passed groves of baobab trees, plump but without leaves because it was still the dry season. We arrived in Mbour and walked the sandy streets of his old neighborhood.

The sun was bright and the skies clear. Not much dust in recent weeks, especially compared to last year. We walked to the beach where his uncle's family used to live. Goats wandered between the pirogues as the surf washed toward the ruins of the house. The night the waves came generated so many questions. "Many years later, I had the answer: climate change," he told me. Rising waters stole land and homes here, as these waves have done all along the West African coast. It can feel overwhelming, the rising seas, the rising temperatures, the accelerating pace of it all.

But there's something we can do, he said, glancing toward the western horizon, toward the Atlantic hurricane nursery. We can take more measurements, document more dust events. With this information, we can improve our computer models. With more precise models, we'll have better predictions about those forces of nature. We'll have more time to adapt. Those waves of change? They won't come as surprises in the night, he said. We'll know they're coming.

---

After a week in Senegal, we were still missing a big orange piece of the Saharan dust puzzle. And we wouldn't find it in Dakar. Senegal isn't really in the Sahara. It's just south of it. We hadn't seen the origin of the dust that affects our hurricanes, and to do that, we would have to travel to Mauritania to the north. The State Department had warned against doing so because it's a military state with terrorist activity. But our Senegal partner, Borso Tall, was confident she could get us across the Senegal River and into Mauritania.

We decided to cross the border from Rosso, a picture of motion and colors as boat operators herded passengers and their cargo into pirogues. Borso somehow found a boat and paid various fees to immigration officials and boat captains, and after an hour, we were in a pirogue. On the other side of the river, Mauritanian officials directed us to a dirt area under a tent. Borso said she overheard an undercover officer talk about our arrival. Immigration officials in flowing white robes interviewed us separately, but my high school French was practically useless, so I mainly nodded. The nodding apparently was good enough, and we made it through, only to have the undercover officer volunteer to be our driver.

Borso came to the rescue again, urging us to slip into a convenience store to dodge the guy. Soon we were with another driver in a beat-up Mercedes, rolling into the Sahara. The landscape changed dramatically in minutes, from beige to the Sahara's rich orange. Camels munched on trees along the side of the road, and one crossed, prompting us to slow down and wait. I could see Whitaker was a little nervous, so I joked, "Why did the camel cross the road?"

No answer from Whitaker.

"Because it was hump day." Still no response.

Soon we were back on our way, and the driver's radio began playing Stevie Wonder's "Part-time Lover."

*Call up, ring once, hang up the phone*
*To let me know you made it home*
*Don't want nothing to be wrong with part-time lover.*

We eventually got out and climbed some orange dunes. I felt the heat coming through the soles of my running shoes. There was no wind, no dust in the air. Nothing to smother a gathering tropical storm.

But, there on top of the orange dunes, it was easy to imagine that someday a gust would come and send dust into the trade winds and toward home.

Two residents in North Carolina's Yancey County look at the devastation on the North Toe River from Helene. Photo by Andrew Whitaker of *The Post and Courier*.

# RISING WATERS

FORECASTS FOR THE 2024 HURRICANE SEASON ARRIVED LIKE distant thunder. Much busier than usual, the meteorologists predicted. Maybe a record-breaker.

They cited several reasons, but one stood out. The ocean is a big battery that soaks up heat energy from the air, and that battery was fully charged. Sea surface temperatures had been rising for years, but in March, a severe heat wave seared West Africa and spilled into the Atlantic's hurricane nursery. By June, ocean temperatures in the tropical Atlantic had reached highs typically seen in mid-August. On cue, thunderstorms that month off West Africa spun into tropical storm Beryl. The storm marched across the Atlantic, plowed into the Windward Islands, and strengthened even more as it moved deeper into the Caribbean. For a time, Beryl's winds topped 160 mph, the earliest Category 5 storm ever observed in the Atlantic. Scientists began talking about whether it made sense to add a Category 6.

Then Hurricane Debby spun to life in late July, striking Florida and drenching parts of South Carolina with two feet of rain. Hurricane Ernesto formed in August, nicked Puerto Rico as it turned north, and slammed into Bermuda.

Then nothing. No tropical storms at all. You could feel the anxiety levels in Charleston drain a bit.

Had forecasters missed something?

My mind moved back to the Sahara Desert and our journey there a few months before. Had dust in the Saharan Air Layer smothered the storms? I looked for clues in satellite images.

No, not much dust at all. If anything, the lack of dust should have allowed more storms to form, not fewer. Had I gotten the Saharan dust story wrong?

I called Gregory Jenkins, the Penn State climate scientist who helped me with our Sahara dust piece.

"What's happening is truly remarkable," he told me. It wasn't the dust, or lack of it. "Something is going on with the African jet."

The African jet stream is a river in the sky that flows east to west, the opposite direction of the North American jet stream. It steers the seasonal monsoons through the Sahel, the semi-arid belt across Africa where the Sahara meets the tropics. But the African jet had twisted sharply north, pulling monsoons into the Sahara itself.

Suddenly, one of the driest places on the planet was awash. Orange floodwaters coursed through the dunes. From Sudan to Morocco, new lakes formed; vegetation began to grow. One area received five years of rain in a month. After drenching the Sahara, these storms exited the African continent off Morocco, instead of Senegal to the south. The Atlantic is much cooler off Morocco. Less heat meant less energy for hurricanes, and the storms fell apart.

"I guess that's good news for Charleston," I said, feeling a little selfish.

"Maybe," Gregory answered. "If the thunderstorms start rolling off Senegal again and have good spin, then it's all back on."

---

Few American cities have suffered as much natural and self-inflicted trauma as Charleston. We've had hurricanes, earthquakes, fires, fevers, and a siege during the Civil War that left much of the city in ruins. Charleston's early prosperity was built on the backs of enslaved people. But denial of slavery's injustice nearly destroyed the city and the country. And yet Charleston today remains a remarkable place despite its tortured history. I'd been seduced long ago by the depth of the city's complexity, its roots tangled in the lessons they offer. I love the city, but my heart is most full when I'm 250 miles from it.

On the Blue Ridge.

I discovered Western North Carolina's mountains in my early twenties after taking my first journalism job in Greenville, South Carolina. Greenville is on the Blue Ridge's eastern flanks, but it's too low to save you from South Carolina's summer sauna. That first summer in the South, I stumbled around in a daze. The air was so thick and heavy. Even in a perfectly air-conditioned room, I sensed the heat outside waiting for me like a bully. Before moving to Greenville, I lived most of my life in cold places like Minnesota, Chicago, and New York. Summer in those cities was your chance to be outside. Now I was confined. I felt cheated, and to escape, I headed to the hills.

I drove my rusty old Saab west, first to the mountain town of Saluda on the North Carolina border, and then higher. In the cooler elevations of Pisgah National Forest, I found spectacular waterfalls. Off the Blue Ridge Parkway, I hiked through trails framed by orange flame azaleas and wild blueberry bushes. In the Smokies, I climbed through green tunnels that opened onto rock faces. On these peaks at sunset, you could see waves of bluish-gray mountains rolling toward the horizon. Over time, I came to terms with South Carolina's heat. But the mountains became even more important to me. Any journalist absorbs a lot of psychological gunk: crooked politicians and business moguls, horrific crimes, failures by people in positions of power to address real problems. The best journalists I've known are furious about these failures. But holding rage has long-term risks. Cynicism and burnout are dangers. So, year after year, I fled to the mountains as a form of therapy, a way of balancing gunk with beauty. Over time, I've gravitated to the mountain's highest peaks: Black Balsam Knob, the stunning white quartz peak in Shining Rock Wilderness Area, the pinnacles over Linville Gorge, and the pitches up Mount Mitchell. Charleston is home, but those old peaks are my sanctuaries.

I'm not alone. People in Florida and other coastal areas in the South also feel the mountains' pull, and not just because it's cooler in the summer. Real estate agents in Asheville began selling the area as a climate change haven. Come to the mountains, where you don't have to worry about hurricane categories!

The range's distance from the coast seemed to guarantee its safety, as well as its elevation and climate. The Southern Appalachians once were as tall as the Western Rockies and European Alps, but the weathering forces of time softened the Blue Ridge into what it is today: a rumpled quilt of peaks and valleys, green in summer and tinted orange and yellow in the fall. Shrouded often in fog, these mountains are a Goldilocks rain forest, not too hot, not too cold. A safe place to weather a storm. A haven.

Then came Helene.

Flooded rivers in South Carolina's Pickens County after Helene.
Photo by Andrew Whitaker of *The Post and Courier*.

The lull in the 2024 hurricane season ended. The record warm waters off West Africa and the Gulf of Mexico launched one storm after another: Francine, Gordon, and Helene. Helene's spinning mass of vapor decimated Florida's Gulf with 140-mph winds when it came ashore. It marched inland through Georgia and toward the mountains. Its spiraling arms extended 420 miles. Helene's swirling storm bands met those old Appalachian peaks. Cooler air over the mountains turned them into catch basins. Staggering amounts of rain fell. Nearly two feet on Mount Mitchell, twenty inches in Spruce Pine. Water surged down the mountains' folds. Streams suddenly became rivers, rivers became lakes, and, in a matter of days, this refuge for so many became a killing zone.

Brown torrents rushed through the Great Smoky Mountains into the Pigeon River, which merges with the French Broad, believed to be among the oldest rivers in the world. From a different direction, the roiling Swannanoa also poured into the French Broad, submerging Biltmore Village near Asheville under twenty-six feet of water. The French Broad rose so high it turned houses into boats. It hadn't been this bad since 1916, known in the area as "the flood to end all floods." It crested at fifteen feet above flood stage, a foot and a half higher than in 1916.

Near Asheville, two grandparents and their seven-year-old grandson were on a roof waiting to be rescued when the building collapsed into the roiling brown Swannanoa. In Marshall, an elderly man clung to a tree until the rising waters of the French Broad swept him away. Helene shredded one mountain town after another, leaving behind mud, rubble, and memories of what once was: Chimney Rock, the folksy tourist stop near Lake Lure; Marshall, an old town along the French Broad; Black Mountain, once voted on TripAdvisor as "America's Prettiest Small Town."

Flooding was less severe in the South Carolina and Georgia Piedmont. But high winds turned trees into wrecking balls. More than 215 people died across the Southeast, with sixty-one lost in Asheville's Buncombe County alone. At least forty-six people died in South Carolina, eleven more than Hurricane Hugo's death toll in 1989. By some estimates, more than 40 trillion gallons of rain fell on the Southeastern United States—enough water to create a lake 3.5-feet deep and the size of North Carolina.

Some storms are so shocking and destructive that they create markers in time. Hurricane Hugo did that to Charleston in 1989. Residents who went through Hugo still talk about events by saying "before Hugo" or "after Hugo." Will Helene create its own marker? I'm not sure. It had a higher death toll and caused more damage than Hugo. But the world is stacking up so many markers. After Helene, wildfires decimated Southern California and Texas, flooding rains struck Ireland and Spain. Humanity has always faced weather catastrophes, but climate change is a force multiplier. Catastrophes punch harder now and more frequently. I'd worried all summer that Charleston would get hammered, only to see one of my favorite places away from home get hit instead. Helene may not be a marker, but it was a reminder: no place, not even a mountain range hundreds of miles from the coast, is safe from something as powerful as a supercharged climate.

---

Denial is a normal reaction to loss and fear, so it makes sense that many people deny that human beings changed the climate.

But the invisible snakes of global warming aren't invisible anymore. They're as real as the saltwater our cars splash through during a sunny-day flood.

Apathy and cynicism are denial's twin cousins. All lead to paralysis, and that's potentially fatal when you're dealing with poisonous snakes or a rapidly warming atmosphere and ocean.

Readers rightly ask me, "But, what can I do?" It's an important question. The forces at work seem overwhelming. The good news is that the solutions aren't invisible snakes. They've been out in the open for decades: More solar and wind power, more nuclear. No more coal and gas. More electric cars and trucks; get rid of those damn gas-powered leaf blowers; replace your oil-burning furnace with a heat pump; eat less meat; vote for leaders who have the courage and honesty to collectively make this stuff happen. So many things, large and small—and they add up. Proof? The United Kingdom has done all of the above. In 2024, Great Britain's greenhouse gas emissions fell to the lowest level since 1872.

The solutions are there, but moving toward those solutions requires focus and willpower.

Beauty helps. Beauty is a reminder that action is worth the effort. When I feel that poison of cynicism and fatalism creeping into my bloodstream, I remember the beauty I've experienced during my reporting: the microscopic jewels in the ocean pumping out our oxygen; tiny birds scampering ghost-like through a sea of grass; a flock of roseate spoonbills lifting off oak trees in the Santee Delta; Greenland's spectacular white and blue ice canyons; sunsets on the ancient peaks of the Blue Ridge; and the wondrous connectivity of it all.

An iceberg off Ilulissat, Greenland, above, and a boneyard beach at Botany Bay.
Photos by Lauren Petracca of *The Post and Courier*.

# WHY THIS BOOK? AND THANK-YOUS

WHY WOULD A LOCAL NEWSPAPER IN CHARLESTON SEND A reporter to Greenland, the Sahara Desert, Bermuda, New Orleans, not to mention Lowcountry swamps filled with copperheads and gators?

My answer: how could it not?

A rapidly warming planet is rapidly reshaping this old city. You can see it in the billion-dollar proposals to wrap the peninsula in a seawall. And in the applications to raise historic structures. And in the decision of a large hospital to move from a flood-prone area downtown to higher ground in another city. And in the moving trucks outside my neighbors' homes, neighbors who felt they also needed to move to higher ground. What's a bigger news story than the fate of your city?

These stories have big stakes, but they often involve incremental changes. And when things happen in small bits, it's easy for news organizations to focus on seemingly more pressing matters, like a new restaurant opening or a trending animal story. Fortunately, *The Post and Courier's* leadership has taken the long view and given reporters the resources to tell complex stories, even ones that take us far beyond South Carolina's borders. This support starts with local ownership, and the Manigault family has steered the paper for generations, now under Pierre Manigault's steady helm. It takes smart publishers, including Ivan V. "Andy" Anderson, Larry Tarleton, Bill Hawkins, and P.J. Browning, to balance a newspaper's commitment to its community and the high costs of doing so.

As with many news organizations, *The Post and Courier* had to find new funding sources as traditional advertising streams dried up. Several nonprofit approaches are filling that void. Our climate coverage received a boost when my flood-weary neighbor, Susan Lyons, donated a large sum to start our Rising Waters Lab. Susan is a retired journalist, and she contributed to honor her partner, Mark Bloom, also a journalist, who died in 2022. Thanks to Susan's generosity, the newspaper created its Rising Waters Lab, which funds climate reporting by two journalists. A portion of every sale will go towards the Rising Waters Lab. (To donate, visit postandcourier.com/rising-waters.) My work also has been supported by the newspaper's nonprofit Public Service and Investigative Fund, another way readers can ensure in-depth local reporting survives these turbulent times. (Donate at postandcourierfund.com.)

The nonprofit Pulitzer Center in Washington, D.C., has been another important partner in our most ambitious journalism projects. (Donate at pulitzercenter.org.) In addition to our Greenland and Sahara projects, the Pulitzer Center supported our Rising Waters coverage in 2020, which was a finalist for a Pulitzer Prize (no relation with the Pulitzer Center, despite the name). Among our stories that year was an effort to test floodwaters for bacteria. Here's some news you can use: Don't wade in the stuff in bare feet. That water is nasty and could make you sick.

In a media carnival of Youtubers and influencers, it's helpful to remember that newspapers are different. A newspaper reporter's story isn't just his or hers. Every story is vetted and shaped by editors. This layer of editing makes newspaper stories more trustworthy than a random dude on Instagram. I've been fortunate to work with many great editors who gave green lights to some of my crazier ideas, including Ed Dawson and Wayne Roper, when I was with the *Greenville Piedmont* back when the first *Back to the Future* movie came out (1989). Over the past thirty years, I worked with talented editors at *The Post and Courier*, including John Burbage, Rick Nelson, Steve Mullins, Barbara Williams, John Huff, Steve Knickmeyer, Mitch Pugh, Autumn Phillips, and especially Doug Pardue and Glenn Smith, two world class project editors who had significant roles in many of these stories. Glenn and I co-wrote the story about the Santee Delta, and he even wrote a song for that project. I'm leaving out many other editors who also set me straight, and I apologize for this lapse. That said, editors understand the need to cut copy. Two amazing photographers, Lauren Petracca and Andrew Whitaker, accompanied me on three of these stories. Their artistry is impressive, as well as their patience with my dad jokes.

I owe a special thanks to the scientists and conservationists featured in this book, especially Dana Beach, Richard Porcher, Chris Crolley and Norm Levine. Thanks also to Elise Lusk and Jacob Hollifield of Evening Post Books for making this project happen, and Paul Middleton of Evening Post Publishing for his years of counsel.

When I'm writing, I often have family members in mind: Margaret Bartelme, Luke, Ava Duryee, Max Duryee, Nicki, Pete, Rachael, Kaia, Maia, and my closest reader, Annie. Thank you for reading, and for your support and love.

# ABOUT THE AUTHOR

TONY BARTELME, A FOUR-TIME FINALIST for the Pulitzer Prize, is a senior projects reporter for *The Post and Courier* in Charleston, South Carolina. Over the past thirty years, his investigative work has exposed government corruption and explored diverse issues ranging from changes in ocean plankton to the global shortage of doctors.

He has received the highest honors in journalism, including recent awards from the Gerald Loeb Foundation, Scripps Howard Foundation, Knight Science Journalism Program at M.I.T., American Geophysical Union, AAAS and Sigma Delta Chi. The S.C. Press Association has twice named him the state's "Journalist of the Year." In 2021, Columbia Journalism School awarded him the John Chancellor Award for Excellence in Journalism, which recognizes a journalist for cumulative achievements.

Tony is the author or co-author of several books, including *A Surgeon in the Village: An American Doctor Teaches Brain Surgery in Africa*. He was awarded a Harvard Nieman Fellowship in 2010 and is a graduate of Northwestern University's Medill School of Journalism.

# ABOUT THE
# PHOTOGRAPHERS

RICHARD VEVERS is the Founder of The Ocean Agency, a Fellow of The Explorers Club, and best known for his leading role in the Emmy Award-winning documentary *Chasing Coral* on Netflix. His work has been featured in numerous publications and documentaries.

JARED BRAMBLETT is a photographer and engineer based in Charleston, SC who bridges civil engineering and visual story telling. A graduate of the University of South Carolina, where he studied Civil & Environmental Engineering, Jared's professional focus lies in water resources and flood resilience while his photographic work centers on a lifelong fascination between humanity and nature.

CHRIS HANCLOSKY is an award-winning video and photo director with over fifteen years of extensive experience in commercial, film, news, studio, event, and sports videography and photography. Chris is a Distinguished Alumnus Award recipient from the University of South Carolina School of Journalism and Mass Communications and has won dozens of awards for Journalism, including video production on a Pulitzer Prize for Public Service for a *Post and Courier* series on domestic violence called "Till Death Do Us Part."

**WADE SPEES** photographed people and events for *The Post and Courier / Evening Post* newspapers from 1980 to 2019 — along with a handful of stories for *The New York Times* each year for much of that time. His thinking that news photography was "All about the subject" kept things fresh for him — and more importantly, he hoped, for the subjects and readers.

**LAUREN PETRACCA** is a freelance photojournalist based in Rochester, NY, working with national editorial and nonprofit clients. A former staff photographer at *The Post and Courier* (2018–2021), she focuses on stories at the intersection of environmental justice and climate change.

**ANDREW WHITAKER** is a staff photographer at *The Post and Courier*, focusing on environmental storytelling. Alongside reporter Tony Bartelme, he reported from Senegal on how Saharan dust affects Atlantic hurricanes. A Michigan native and Central Michigan University graduate, he joined the paper in 2018 and also leads storm coverage as a hurricane chaser.

# NOTES ON SOURCES

## SNAKES

Material in this chapter was inspired in part by my previous book, *A Surgeon in the Village*, a nonfiction account of Dr. Dilan Ellegala's effort to teach brain surgery in Tanzania. I spent lots of time with neurosurgeons and learned some things about the brain. Like the amygdala – they're a pair of almond-shaped structures in the brain that would light up if we suddenly encountered a snake at our feet. I found additional inspiration in David Eagleman's excellent book, *Incognito: The Secret Lives of the Brain*. One of his key hypotheses is that each one of us has a symphony of software-like programs running in our brains, and that these programs are constantly competing for primacy. Daniel Gilbert's book, *Stumbling on Happiness*, is where I first saw his description of the brain as a "great get-out-of-the-way machine."

The invisible snake metaphor is grounded in real experience. During our reporting in 2019 on the Santee Delta, I was on a sunny trail with Glenn Smith and Lauren Petracca, when Glenn almost stumbled on a cottonmouth.

## CHASING CARBON

Most stories begin with a conversation. In 2016, David Quick, a colleague at the newspaper at the time, mentioned that he'd heard about a special camera that could "see" carbon dioxide. With that seed planted, I corralled the camera from FLIR for an article in *The Post and Courier* published Oct. 22, 2016, under the same title and found here: https://data.postandcourier.com/saga/on-the-edge/page/3. Links embedded in that story take you to studies and datasets cited in the article and this chapter. I updated some of the carbon dioxide statistics gathered by NOAA at its station in Hawaii. The latest carbon dioxide readings can be found here: https://gml.noaa.gov/ccgg/trends/. In mid-2025, they exceeded 428 parts per million, a record high and a 50 percent increase since pre-industrial times.

My conversation with Lonnie Carter, former chief executive officer of Santee Cooper, was discussed in a *Post and Courier* article published Nov. 7, 2007, with the headline: "Coal debate swirls around Santee Cooper."

A number of scientists helped me shape the original story and this chapter, including Mitchell Colgan of the College of Charleston; James T Morris, director, Belle W. Baruch Institute for Marine and Coastal Sciences, University of South Carolina; and Mark Boccella, FLIR Americas Business Development Manager, Optical Gas Imaging.

## EVERY OTHER BREATH

This chapter draws largely from an article in *The Post and Courier* by the same name and published March 17, 2016. Links to sources and studies are embedded in the story found here: https://data.postandcourier.com/saga/plankton/longread.

NOTES ON SOURCES

In addition to those named in this chapter, a number of other local and nationally recognized scientists contributed thoughts and information about this complex issue. Jack DiTullio, an assistant professor of biological and chemical oceanography at the College of Charleston, explained the carbon cycle and other notable breakthroughs in phytoplankton research. Ryan Rykaczewski, a biologist at the University of South Carolina's Marine Science Program, described the history of the deep scattering layer and other interesting behaviors of zooplankton. Samuel Laney, a biologist at the Woods Hole Oceanographic Institution in Massachusetts, gave an excellent Plankton 101 lesson.

In Bermuda, Becky Garley, Afonso Goncalves, Sam Stevens and Tim Noyes allowed me to observe their sampling by BIOS for their Bermuda Atlantic Time Series, better known in science circles as (BATS). Other scientists consulted include: Dianne Greenfield, a phytoplankton expert and associate professor with Belle W. Baruch Institute for Marine and Coastal Sciences, University of South Carolina; Tammy Richardson, a biology professor at USC; Samantha de Putron, an expert in corals at the Bermuda Institute of Ocean Sciences (BIOS); Bill Curry, chief executive officer of BIOS; and Leocadio Blanco-Bercial, a BIOS zooplankton expert.

Dennis Allen has retired as resident director of the Belle W. Baruch Marine Field Laboratory and is now a Distinguished Professor Emeritus at the University of South Carolina.

## FADE TO WHITE

This chapter is based on an article with the same name published in *The Post and Courier* on Oct. 19, 2016. Links to cited studies and sources can be found here: https://data.postandcourier.com/saga/on-the-edge/page/2.

A number of other scientists and others not quoted in the chapter also were consulted, including: Peter Moeller, research chemist with the National Oceanic and Atmospheric Administration and Medical University of South Carolina; Cheryl Woodley of the College of Charleston and NOAA Hollings Marine Laboratory; Ray Swaggerty, who has done extensive research on the Frederick W. Day; and Samantha de Putron, Bermuda Institute of Ocean Sciences.

## SCUM

This chapter is based in part on two stories in *The Post and Courier*, "Scum" and "Toxic Algae Blooms: A growing worldwide menace." Both were published Sept. 17, 2017. Links to sources cited in the chapter can be found here: https://data.postandcourier.com/saga/algae-sunday/page/2.

## LOWCOUNTRY ON THE EDGE

This chapter is a gumbo of stories in *The Post and Courier*, including "Lowcountry on the Edge" (Oct. 2, 2016), "Lessons of Louisiana" (Dec. 31, 2017), and new material based on my observations of the American Geophysical Union conference in New Orleans in 2017. Data and study citations are embedded in links in these stories: https://data.postandcourier.com/saga/on-the-edge/page/1 and https://tinyurl.com/ms6ruyv6.

## INTO THE GULF STREAM

This chapter is based on stories *The Post and Courier* published Sept. 5, 2018. Data sources are embedded in links in stories here: https://www.postandcourier.com/gulfstream. At the time, the report was among the first nationally by a general media outlet to describe

NOTES ON SOURCES

the vulnerability of the Atlantic's currents to a rapidly warming climate. This chapter is based on more than thirty interviews with scientists across the world and supplemented by research papers, documents generated by the Ben Franklin expedition, historical accounts and computer simulations on an open-source program developed by NOAA. Gene Feldman, a NASA oceanographer, shared a storehouse of photos and documents he's assembled over the years. From his home in Florida, Don Kazimir showed me logs and other primary sources. Some descriptions of the mission come from Jacques Piccard's book, *The Sun Beneath the Sea*. Others were inspired by Hans Leip's 1957 book about the Gulf Stream, *River in the Sea* and Stan Ulanski's 2008 book, *The Gulf Stream*.

In addition to Harry L. Bryden, William Johns and Tom Rossby, I consulted other leading Gulf Stream scientists, including: Susan Lozier of Duke University; Robert Todd of Woods Hole Oceanographic Institution; Tal Ezer of Old Dominion University; Vincent Saba, a NOAA research fishery biologist at Princeton University; William Sweet, a NOAA oceanographer in Florida; and Dana Savidge of University of Georgia's Skidaway Institute of Oceanography.

Matthew Upton, president of Roffer's Ocean Fishing Forecasting Service, and Jenifer Clark, a professional meteorologist known as the "Gulfstream Lady," also provided key insights. Doug Helton, NOAA Emergency Response Division, pointed us toward oil spill response simulations for old shipwrecks.

And the Hurricane Fleet in Calabash, N.C., took us into the Gulf Stream itself.

## THE GREENLAND CONNECTION

So many things about Greenland are big—except for its population. Roughly 56,000 people live here, and most towns are reachable only by boat or plane—transportation that generates a large carbon footprint. We offset this reporting with carbon credit purchases from Terrapass, which funds projects to reduce carbon dioxide emissions around the world.

Much of this story is drawn from our project for *The Post and Courier* in 2021. Links to studies are embedded in the newspaper's story here: https://tinyurl.com/2jmxzsf3.

In Greenland, Lauren Petracca and I navigated iceberg-flecked waters and fought off a suicidal drone with Claus Foss Hansen, who operates WhaleTours in Ilulissat. Karen Buus and Arne G. Petersen of AirZafari showed us Greenland from above. Ringo Mathiassen, a fisherman in Ilulissat, showed us his sled dogs and helped us understand how climate change affects people in the Arctic. Jim Haffey of Kenn Borek Air gave us an incredibly thorough briefing before our flight with NASA scientists. ("If the plane crashes and everyone else is unconscious, here's how you turn off the engines.")

In addition to the scientists quoted in the story, we interviewed more than a dozen other researchers. Special thanks to Eric Rignot, Eric Larour, Mike Wood, and Jane Lee at NASA, Natalya Gomez from McGill University in Canada. Christian Rodehacke of the Danish Meteorological Institute helped with melting calculations. And Susan Joy Hassol helped connect us with key researchers in the project's early stages.

## GHOST BIRD

Reporting for this chapter is based on an article with the same title published Sept. 10, 2020, in *The Post and Courier*. Citation sources are embedded in links in the story here: https://tinyurl.com/4vmxs9ut.

With advice from black rail researchers, we published the original story at the end of the black rail's breeding season when they're much less vulnerable to human interference. Christy Hand, then a biologist with the South Carolina Department of Natural Resources, was kind enough to organize excursions into South Carolina marshes, even during the pandemic. Bryan Watts, director of The Center for Conservation Biology, has been working on eastern black rail issues since the late-1980s and helped provide important context. Members of the Carolina Bird Club also contributed their experiences, including Lewis Burke and Craig Watson.

In addition to those quoted in the story, thanks also to Erin Weeks and Jamie Dozier at the Department of Natural Resources, and Stephanie Kurose of the Center for Biological Diversity. Ernie Wiggers and Beau Bauer at Nemours Wildlife Foundation also described their work to protect land for black rails. After the story in *The Post and Courier* was published, the federal U.S. Fish and Wildlife Service listed the eastern black rail as "threatened" under the Endangered Species Act, giving the rare bird added protections.

## OUR SECRET DELTA

This chapter is rooted in a story we produced for *The Post and Courier* on Sept. 20, 2019. I worked closely with Glenn Smith and Lauren Petrracca on the project. We began our reporting in the summer, and at first, we thought this timing was a mistake. The heat and humidity were staggering. We found ourselves steps away from cottonmouths and alligators. We plunged into pluff mud and tromped through poison ivy. We were covered with mosquitoes and bitten by ticks and deer flies. Uncomfortable, yes. But in the end, it gave us a more realistic understanding of the brutal climate enslaved people faced in the swamps and rice fields.

We interviewed more than forty people, and we were fortunate to have patient guides, including botanist Richard Porcher. Selden B. "Bud" Hill, founder of the Village Museum in McClellanville, and Randy McClure, its current director, also helped frame the region's evolution, as did Vennie Deas Moore, a cultural historian who offers tours at Georgetown's Rice Museum; Michael Prevost, who has worked for decades on land conservation projects; and William Garrett, a retired guide who worked for years leading hunts at the Santee Gun Club. Many scientists and researchers also helped educate us, including Raymond Torres, John Hodge and Brent Fortenberry. Achi Treptow, a South Carolina Department of Natural Resources biologist and head of the Upper Coastal Waterfowl Project, organized a revealing visit of Murphy Island, which many in the delta say is among the buggiest places they've experienced. (They're correct.) Jamie Dozier, a state wildlife biologist and project leader of the Tom Yawkey Center, also offered us deep looks at this important public refuge.

We took two trips with Coastal Expeditions, one to Crow Island and a second to Cedar Island. Guides Chris Crolley and Gates Roll have deep respect and knowledge of the delta and offer their observations with refreshing lyricism.

We leaned on several helpful books, including Porcher's *Market Preparation of Carolina Rice* and *Rice to Ruin* by Roy Williams III and Alexander Lucas Lofton, both published by University of South Carolina Press.

We also drew from Matthew Allen Lockhart's excellent thesis, "From Rice Fields to Duck Marshes," especially sections that describe the Santee Gun Club's formation and the tension between locals and the "second Yankee invasion." Judith Carney's *Black Rice* (Harvard University Press) also was a helpful resource.

NOTES ON SOURCES

## THE SAHARA CONNECTION

Stories are often like links in a chain. After we finished "The Greenland Connection," I'd begun thinking about other ways to connect global climate patterns to the South Carolina Lowcountry. Thanks to a Pulitzer Center travel grant, we traveled to Senegal and Mauritania in 2023 for several stories under "The Sahara Connection" umbrella. *The Post and Courier* published our story about the desert's dust on June 23, 2023. Soon after, we also published reports about a Senegal city's effort to build a seawall in a project we called "Lessons of Senegal." All of those stories and source citations can be found here: https://www.postandcourier.com/sahara/.

Before we left for West Africa, Michael Kaplan, professor emeritus of atmospheric sciences at the Desert Research Institute in Nevada, provided important background and guidance, as did Joseph Prospero, professor emeritus at the University of Miami's Rosenstiel School of Atmospheric Sciences. Prospero was quick to note the early and under-recognized Saharan Air Layer discoveries of other scientists, such as Guillermo Luloaga of Venezuela and Christian Junge. Jason Dunion, a researcher with the NOAA's Hurricane Research Division, also discussed his important dust transport work.

For this story, we traveled into Mauritania to set our eyes on the Sahara itself, a potentially perilous journey in a pirogue across the Senegal River to a country that doesn't see many foreign journalists. Borso Tall, a journalist based in Senegal, was instrumental in making that happen, as well as acting as a French and Wolof translator for many of our interviews. Amadou Gaye, a professor at University Cheikh Anta Diop of Dakar, was a fountain of experience who introduced us to many of his fellow climate researchers. Likewise, Abdoulaye Deme and Boubou Sy at Gaston Berger University in Saint-Louis, Senegal, were important local sources.

Senegal has had one of the most stable democracies in West Africa, but just before our visit and near the end, the country had deadly protests and riots. Add political instability to the already daunting number of hurdles West African scientists face in getting their research done and distributed. Almamy Badiane, a guide and translator in Dakar, also helped us gain a deeper understanding of Senegalese culture. Flying creates a large carbon footprint, and we offset this trip with carbon credits from Terrapass.

## RISING WATERS

In 2024, I worked on another Pulitzer Center-funded story with Borso Tall, our journalism partner in Senegal. We called the project "In Hot Water." It detailed how a rapidly warming climate had heated the cauldron in the Atlantic Ocean off West Africa that generates many of our worst hurricanes. I drew some of the chapter's material from that story, which can be found here: https://pulitzercenter.org/stories/hot-water.

Kerry Emanuel, the noted climate scientist from Massachusetts Institute of Technology, isn't quoted in this chapter, but his counsel in shaping my reporting over the years has been invaluable.

During my career, I've covered at least seventeen hurricanes that affected South Carolina in one way or another, and 2024's hurricane season stood out for its dramatic cease fire and then Helene's devastating march through Florida, Georgia, and the Carolinas. Some details about Helene in this chapter were drawn from a report I co-wrote with Glenn Smith called "Deadly Helene," which was published Oct. 3, 2024, in *The Post and Courier*.

## NOTES ON SOURCES

The source for Great Britain's carbon dioxide emissions is Carbon Brief's analysis in March 2025, which also showed that country's use of coal had dropped to the lowest level since 1666.

# ALSO BY TONY BARTELME

A SURGEON IN THE VILLAGE

SECOND CHANCE: THE MARK SANFORD STORY

**THE BRIDGE BUILDERS**
with Jessica Vanegeren

**INTO THE WIND**
with Brian Hicks

www.ingramcontent.com/pod-product-compliance
Lightning Source LLC
Chambersburg PA
CBHW020848160426
43192CB00007B/834